SpringerBriefs in Applied Sciences and Technology

# PoliMI SpringerBriefs

AF173103

**Series Editors**

Barbara Pernici, DEIB, Politecnico di Milano, Milano, Italy

Stefano Della Torre, DABC, Politecnico di Milano, Milano, Italy

Bianca M. Colosimo, DMEC, Politecnico di Milano, Milano, Italy

Tiziano Faravelli, DCHEM, Politecnico di Milano, Milano, Italy

Roberto Paolucci, DICA, Politecnico di Milano, Milano, Italy

Silvia Piardi, Design, Politecnico di Milano, Milano, Italy

Gabriele Pasqui⬤, DASTU, Politecnico di Milano, Milano, Italy

Springer, in cooperation with Politecnico di Milano, publishes the PoliMI Springer-Briefs, concise summaries of cutting-edge research and practical applications across a wide spectrum of fields. Featuring compact volumes of 50 to 125 (150 as a maximum) pages, the series covers a range of contents from professional to academic in the following research areas carried out at Politecnico:

- Aerospace Engineering
- Bioengineering
- Electrical Engineering
- Energy and Nuclear Science and Technology
- Environmental and Infrastructure Engineering
- Industrial Chemistry and Chemical Engineering
- Information Technology
- Management, Economics and Industrial Engineering
- Materials Engineering
- Mathematical Models and Methods in Engineering
- Mechanical Engineering
- Structural Seismic and Geotechnical Engineering
- Built Environment and Construction Engineering
- Physics
- Design and Technologies
- Urban Planning, Design, and Policy

Chris Hesselbein · Paolo Bory

# Infrastructures of Reality: Metaverse Stories, Spaces, Bodies

POLITECNICO
MILANO 1863

Chris Hesselbein 🆔
Department of Management Engineering
Politecnico di Milano
Milan, Italy

Paolo Bory 🆔
Department of Design
Politecnico di Milano
Milan, Italy

ISSN 2191-530X          ISSN 2191-5318  (electronic)
SpringerBriefs in Applied Sciences and Technology
ISSN 2282-2577          ISSN 2282-2585  (electronic)
PoliMI SpringerBriefs
ISBN 978-3-031-97166-2     ISBN 978-3-031-97167-9  (eBook)
https://doi.org/10.1007/978-3-031-97167-9

This work was supported by Politecnico di Milano.

This Springer imprint is published by the registered company Springer Nature Switzerland AG
The registered company address is: Gewerbestrasse 11, 6330 Cham, Switzerland

If disposing of this product, please recycle the paper.

# Preface and Acknowledgements

This book is the result of an intense process of dialogue and translation between our respective disciplinary backgrounds and personal predilections, but it is also the outcome of mutual curiosity and a good friendship. Our journey into the 'metaverse' was not a lonely one; we would like to thank all the colleagues, friends, and family members who accompanied us on this challenging, often confusing, but nonetheless exciting path.

We are grateful to the colleagues and friends of the original, 'real' Meta—the interdisciplinary research unit in the social sciences and humanities founded at the Politecnico di Milano in 2016—who have stimulated and challenges us in various but always valuable ways. We wish to thank Stefano Canali in particular for thinking through issues of datafication with us. Additional thanks also to Hernán Bobadilla, Sergio Preston, Camilla Rinciari, and Sahar Tavakoli for listening to our ideas, returning them to us in an improved form, and suggesting alternative paths.

Next, we also thank all the participants in the conference panel that we co-organized (with Stefano Canali) at the STS Graz conference in 2023 titled "Understanding the Metaverse: theoretical, empirical, and critical challenges for a new(?) internet age".

Last, we thank all the members of the PRIN 2022 research project "From netizens to metazens: Narratives of virtual worlds and civic engagement in the metaverse", namely Federico Biggio, Tiziano Bonini, Stefano Bory, Fabio Iapaolo, Gianfranco Pecchinenda, Marcus Pingitore, Tonio Savina, and Paolo Volonté. This book is also the result of all the questions we have asked ourselves together in the last two years, but above all this one: "Cce cazzu ede lu metaversu?".

We also wish to acknowledge the following funding support:

Paolo Bory acknowledges financial support under the National Recovery and Resilience Plan (NRRP), Mission 4, Component 2, Investment 1.1, Call for tender No. 104 published on 2.2.2022 by the Italian Ministry of University and Research (MUR), funded by the European Union—NextGenerationEU—Project Title "Exploring narratives of virtual worlds and civic engagement in the Metaverse"—CUP D53D23012780006—Grant Assignment Decree No. 1060 adopted on 17/07/2023 by the Italian Ministry of Ministry of University and Research (MUR).

Chris Hesselbein acknowledges financial support under the National Recovery and Resilience Plan (NRRP), Mission 4 Component 2 Investment 1.3—Call for tender No. 341 of 15/03/2022 of Italian Ministry of University and Research funded by the European Union—NextGenerationEU. Award Number: PE00000004, Concession Decree No. 1551 of 11/10/2022 adopted by the Italian Ministry of University and Research, CUP D73C22001250001, MICS (Made in Italy—Circular and Sustainable).

Milan, Italy                                                                         Chris Hesselbein
                                                                                      Paolo Bory

**Competing Interests** The authors declare no competing interests.

**Ethics Approval** This works does not involve any human or animal participants and has therefore not sought any formal ethics approval.

# Contents

# About the Authors

**Chris Hesselbein** is a researcher in Science and Technology Studies in the Department of Management Engineering at the Politecnico di Milano. His research focuses on how knowledge and technology, both digital and more mundane, are co-constructed with conceptions of social order and self-identity, as well as how the production and consumption of technoscience inform, and often naturalize/ normalize, our understandings of embodiment, materiality, and aesthetics. He is the author of several chapters in edited volumes as well as articles in international journals, such as *Information, Communication and Society*, *Science as Culture*, and *Fashion Theory*. His greatest accomplishments, however, take place in the kitchen.

**Paolo Bory** is a researcher in Sociology of Culture and Communication in the Department of Design at the Politecnico di Milano. His research focuses on the imaginaries and histories of digital media and technologies such as the Internet, AI, MR, and supercomputing. He is the author of the monograph *The Internet Myth. From the Internet Imaginary to Network Ideologies* (University of Westminster Press 2020) and of several articles in international journals, including *Convergence, AI and Society, Information, Communication and Society, New Media and Society*, and *Public Understanding of Science*.

# Chapter 1
# Why the Metaverse?

**Abstract**  This introductory chapter situates the term 'metaverse' in multiple corporate and academic literatures, unpacks its various uses and technological aspects, and makes the case for its continued value as an analytical term. Subsequently, we redefine this as a process of 'metaversification' involving the virtualization, datafication, and infrastructuralization of spaces and bodies. This tripartite process involves the drawing in of various activities into virtual worlds while simultaneously pushing the logics of virtualization and datafication into existing physical spaces and bodily practices. Taken together, these processes position 'tech' companies as new infrastructures of reality.

## 1.1  From the Metaverse to Metaverses

The word 'metaverse' is a portmanteau that betrays an all-encompassing ambition. On the one hand, it denotes a state of being that is above or beyond the present, a transcendence towards something of a higher order (*meta-*). On the other hand, it signifies a new *-verse* or world that encompasses the totality of existing things and beings. Originally a term proposed by sci-fi novelist Neil Stephenson in 1992 to describe a dystopian future in which national governments have ceded their power over public institutions, spaces, and services, to private corporations, now the 'metaverse' has—rather ironically—been taken up by large technology companies across the world to describe their specific visions of the future. Mark Zuckerberg famously changed the company name from Facebook to Meta and declared the metaverse as "the next chapter for the internet" (Meta, 2021). Jensen Huang, CEO of Nvidia, has stated that the metaverse "is where we will create the future" (Shapiro, 2021). Satya Nadella of Microsoft, in a similar vein, has asserted that the metaverse allows "us to embed computing into the real world and to embed the real world into computing", and that this "is not just transforming how we see the world. It's changing how all of us actively participate in it" (Nadella, 2021). Last but not least, the CEO of Apple, Tim Cook, recently heralded the company's entry into spatial computing via the Apple Vision Pro headset as a "new era for computing" that follows in an evolutionary line

© The Author(s) 2025
C. Hesselbein and P. Bory, *Infrastructures of Reality: Metaverse Stories, Spaces, Bodies*,
PoliMI SpringerBriefs, https://doi.org/10.1007/978-3-031-97167-9_1

from the eras of personal and mobile computing that preceded it (Apple, 2023). That is to say, according to heads of the largest technology companies in Silicon Valley and beyond, the metaverse encompasses the future of the world, a new paradigm on par with similarly hyped technologies such as Artificial Intelligence (AI) that promise to disrupt everything and fundamentally transform society ranging from the ways we work and socialize to how we consume and spend our leisure time. Digital dreams are finally taking over physical realities, it seems, again.

Going beyond such grandiose visions and assertions, and looking at the more specific descriptions of the metaverse by technology companies, commentators, and researchers, however, its underpinning technologies, goals, and outcomes diverge widely. For instance, a common interpretation of the metaverse is that it involves novel virtual environments or worlds in which, according to some, virtual reality (VR) headsets are central. Others envision a more blended approach known as augmented reality (AR), which combines the physical world and overlays this with virtual elements. Somewhere in between, one might encounter terms such as mixed reality (MR) and extended reality (XR), which are frequently used as umbrella terms for both VR and AR. Still others, however, claim that the metaverse is already here, namely in the form of virtual worlds and gaming platforms, such as Second Life or Fortnite, neither of which require VR or AR technologies. Moreover, whereas companies such as Microsoft and Nvidia are focusing on software and chip development for enterprises and industries, Meta is focusing on hardware for commercial use, such as its Quest series of head-mounted displays, and on developing its VR platform Horizon Worlds. Meanwhile, game or game-engine developers, such as Epic Games, Roblox, Activision Blizzard, and Unity, have set up enormous gaming platforms or have become central to the production of media content and simulations across multiple industries, including film, architecture, design, and construction. Here, so-called digital twins of large infrastructures—such as airports or cultural heritage sites as well as entire cities or even countries—as well as everyday objects and bodies, are seen as a fundamental part of constructing the metaverse. The overlap between chip makers, such as Nvidia, and the development of virtual environments is particularly intriguing, suggesting that a still under-examined relationship exists between virtuality and the rise of AI. Maybe it is not so coincidental that graphics-processing chips—long used to make games look more realistic—are now being used to reshape reality. What is more, there also appears to be a relationship between AI models and how these are integrated into virtual characters in games as well as robots in the 'real' world, again suggesting a connection between virtuality, AI, and automation. And not to be forgotten, there are also some who claim that decentralized technologies, such as blockchains and cryptocurrencies, are fundamental to the future of the metaverse. Last but not least, besides the visions of technology companies in the Global North, there are artistic renditions of the metaverse that seek to explore this emergent technology for the development of new creative forms and practices, sometimes with the explicit goal of moving beyond or even critiquing Silicon Valley ideologies. And there are also large initiatives in countries beyond the US, such as China, Japan, and Korea—where gaming and virtual practices are more integrated into society and 'virtual influencers' are already more widespread—that are seeking to develop their own versions of the metaverse.

The term metaverse is used to describe vastly different things depending on which social group or corporate actor you look at. Each appears to have its own different yet equally under-defined version, but almost all share a similar sense of ambition in terms of revolutionizing society by means of metaversal technologies and transforming the ways in which people work, interact, and consume. In short, the term can cover almost anything. That is to say, the 'metaverse' has become a buzzword that is regularly bandied about with few real attempts at defining it other than the promise of radical disruption and transformational change. Talk about the metaverse can thus indicate not only a commitment to a certain kind of digital innovation or a specific technological vision of the future, but also be employed as an empty signifier or a marketing term to drive stock prices and attract financial investment. In other instances, the term can be used to shape technological narratives, create cultural hype, and to placate (or confuse) regulators. Its definitional multiplicity thus allows it to be used by various actors according to their specific goals and interests. And even when it is elaborated upon, the definitions tend to vary quite widely, frequently emphasizing either virtual spaces or wearable devices, or some combination of both. In other words, it is difficult to write about the 'metaverse' in a coherent manner because of all these disparate perspectives and developments, some of which appear to overlap in uncertain ways or even contradict each other.

Because of such conceptual vagueness as well as the corporate connotations of the term 'metaverse', many academic researchers have been hesitant to adopt it or have foregone using it at all, instead preferring to focus their analyses on better-defined phenomena, such as VR (Carter & Egliston, 2024; Evans, 2019; Messeri, 2024), AR (Bolter et al., 2021) or gaming and 'microverse' platforms (Evans et al., 2022). Others have trained their focus predominantly on one specific technological case or company. The latter generally being Meta because of its dominant role in shaping discourse around the metaverse as well as the enormous resources it is mobilizing to develop its headsets and virtual platforms. Indeed, some researchers have indicated that they believe that it is not useful, or at the least potentially problematic, to employ the term 'metaverse' (e.g., Carter & Egliston, 2024; Evans et al., 2022). This is in part because the term can reproduce and contribute to corporate hype, but also because of its close association with Meta and its problematic history in terms of privacy regulations, free speech, and datafication. In addition, it is quite possible that the term will not retain its conceptual usefulness—neither for technology companies nor academic researchers—and disappear entirely in the future, to be placed beside other terminological relics of the digital age such as 'cyberspace'. Moreover, the conceptual vagueness of the term is also useful precisely because it presents a future vision that technology companies are ostensibly working towards and that justifies their dominance of this industry while simultaneously obscuring the potential harm of their actions in the present, which are often framed as merely short-term issues (Beer, 2024). This is particularly pertinent to processes of datafication and how the grabbing of personal data is justified for the development of powerful AI models, which, it is argued, will enable drastic improvements in terms of economic growth, individual productivity, and the broader 'knowledge economy'.

The reasons for not adopting a vaguely defined buzzword or overhyped concept are, of course, understandable and justifiable, particularly for academic researchers who would do well to keep a critical distance rather than uncritically using the concepts of corporate actors. However, despite being aware that using such a poorly defined term can be fraught, we nevertheless wish to continue to employ the 'metaverse' as an analytical concept. The word 'metaverse', as mentioned above, denotes an evolution or transcendence towards a higher order that is above or beyond the current state of the world. Although rarely said in so many words, this new state of affairs is a virtualized and datafied one, and above all, largely operated, maintained, and controlled by large corporations and thus predominantly reflecting their commercial interests. Continuing to employ the term 'metaverse' is, we believe, important not only to highlight the origins of the term in sci-fi literature and therefore to remind us of its dystopian connotations, but also to underscore the all-encompassing and totalizing ambitions of the 'Big Tech' companies seeking to reshape the world according to their particular visions, interests, and goals. By employing the term 'metaverse' we wish to remind you of the utopian ambitions of these companies and the enormous power and wealth of resources that underpin them. This is an attempt at reconfiguring our reality.

Despite its clear shortcomings, the term 'metaverse' can also be useful for two additional reasons. First, the history of technology and media has always been shaped by umbrella terms that encompass a vast array of technologies. For instance, 'mass media' has long been used to describe different technologies, such as the printing press, radio, cinema, and television. In the 1990s, 'cyberspace' was used to cover almost all devices, networks, environments, and human and non-human entities involved in the emergence of the World Wide Web. Even now, in the middle of the ballyhoo around AI, this term is used to describe various softwares, systems, models, and applications that range from facial recognition to natural language processing, from image generation to predictive algorithms. Like its terminological predecessors, the metaverse is thus helpful as a semantic shortcut to indicate a range of ongoing computational processes and developments—i.e. the implementation of virtual worlds, big data, AI systems, and robotics—that might appear somewhat disparate but that can be linked together in order to better examine their convergences and future implications. Second, the term 'metaverse' helps us to highlight and critically analyse key features that are common across most if not all corporate metaverse projects, namely the capture and control over more detailed information about spaces and bodies through a process of virtualization, and by consequence, the positioning of the metaverse as an all-encompassing infrastructure for society. Both aspects are inextricably connected, and both are already central to the monopolizing ambitions of 'Big Tech' companies seeking to go beyond the present (*meta-*) and construct a new world (*-verse*).

## 1.2  Virtualization, Datafication, and Infrastructuralization

At the heart of the ambition to build the metaverse lies, we argue, a triple set of processes, namely virtualization, datafication, and infrastructuralization, which, as other scholars have already noted, have long been a core part of the corporate ideology found in Silicon Valley (Barbrook & Cameron, 1996; Boyd & Crawford, 2012; Sadowski, 2019). Virtualization indicates the creation of virtual versions of physical systems, objects, people, or processes, and the subsequent blurring of the boundaries between these physical and virtual versions. Datafication refers to the process of converting all aspects of human life and its environment into digital data for quantified analysis and decision making (Ruckenstein & Schüll, 2017; Van Dijck, 2014). Given the past actions of many of the technology and media companies driving the development of the metaverse—e.g., Meta, Alphabet, Microsoft, Tencent—it is clear that the 'data imperative' (Fourcade & Healy, 2017) lies at the centre of their efforts. Metaverse datafication gives rise to a new and important series of issues and implications on the level of knowledge production, scholarly research, and regulation (Hesselbein et al., 2024b). But the expansion of data collection is only one step, albeit a crucial one. Infrastructuralization is the next step, which is the process of making the socio-economic wellbeing and stability of societies both central to and dependent on the functioning of the platforms, technologies, and services offered by 'Big Tech' companies (Hardaker, 2025; Hendrikse et al., 2022; Kerssens & Van Dijck, 2023; Plantin et al., 2018; Van der Vlist et al., 2024). The dominant vision underpinning the development of the metaverse—as the next big thing' if not 'the future'—suggests a world in which most, if not all, economic, cultural, and social activities are subsumed into an infrastructure in and through which all these activities are mediated and take place (Hesselbein et al., 2024a). The metaverse is thus presented less as a device or platform to occasionally use than a space to occupy as well as work, and to some extent live, in the long term. In other words, an infrastructure that is central to the operation and functioning of society as well as to the expression and cultivation of our various cultural and individual pleasures and pastimes. Needless to say, this infrastructure is designed, constructed, managed, and maintained by the actors at the centre of current metaverse discourse and development. Much like how the commercial internet was ultimately shaped by the visions of the dominant corporate players and their platforms—taking personal data, which is sold to advertisers and data brokers, in exchange for seemingly 'free' services such as search engines and social media platforms (Hwang, 2020)—it is highly likely that the metaverse will, if the dominance of such actors goes unacknowledged and unchallenged, be shaped in a similar manner.

The infrastructural ambitions and totalizing tendencies of 'metaverse' companies in their efforts to capture and reshape spaces and bodies—one might call this the 'metaversification' of society—is not merely a matter of data colonialism (Couldry & Meijas, 2019; Thatcher et al., 2016), platformization (Nieborg & Poell, 2018), or surveillance capitalism (Zuboff, 2019). Although these are important terms that are indeed pertinent to the emergence of the metaverse, particularly in their emphasis on the expansionary and colonialist grabs of personal data by technology companies for commodification, exploitation and surveillance, we believe they need further elaboration. Datafication in and through virtual spaces and devices is not simply a continuation of previous processes of datafication, which mainly involved textual and visual data from online platforms and mobile or connected devices. Instead, metaverse datafication involves a qualitative change in terms of the types of data that are collected, specifically of physical bodies and the environments in and through which they move and live. In short, this is a new, important, and different kind of embodied, spatial, and environmental data that are far more intimate and insightful than previous forms of data. Moreover, these new types of data are likely to be crucial for teaching AI systems how to gain a better understanding of the human bodies they interact with as well as the spatial contexts in which they operate. Similarly, although many virtual worlds can be understood as platforms, particularly in how they seek to position themselves in different economic sectors and spheres of life—and thus drawing in activities once carried out elsewhere as well as shaping cultural practices and imaginations—we believe platformization captures only one important part of the story. The other important part is that the logics of virtualization and datafication are increasingly being pushed beyond platforms and into our environments and bodies.

In addition, we also assert that the metaverse is not simply an example of 'remediation' (Bolter et al., 2021) in which VR or AR media reference, re-represent, incorporate, and compete with previous media such as film and television. Although VR and AR certainly can be understood as a form of 'reality media' in how they mediate our perception of the everyday world and therefore redefine or construct our reality, we believe the emphasis on media—as powerful and important as these are—also does not go far enough in describing the infrastructural ambitions underpinning metaverse development. The metaverse certainly encompasses the production and consumption of 'media' as one of its components—particularly in how new types of 'content' are being produced in gaming worlds, for example—but it also goes far beyond this in the sense that it seeks to draw existing practices into virtual spaces while simultaneously pushing the logics of virtualization and datafication into existing physical spaces and bodily practices. In short, we are not just talking about representations but about spaces in which people work, play, and to some extent live. It is precisely for this reason that we have employed the terms 'infrastructures' and 'reality' in the title of this book rather than 'media' or 'technology'.

We therefore conceptualize metaversal 'infrastructures of reality' in a tripartite manner. First, they encompass companies, platforms, and technologies as well as the services these offer and the datafication processes they involve. Second, they are geared towards physical spaces and human bodies as well as the practices they engage in, such as work and leisure. And these infrastructures also involve the discourses, narratives, and concepts that provide a framework for shaping—at a broad sociocultural level—the understanding, acceptance, and even desire for metaversal development and implementation. Taken together, these processes of 'metaversification' challenge the boundaries between technology, embodiment, and space, reconfiguring each into a new reality.

We believe that three aspects of this reality, namely the virtualization, datafication, and infrastructuralization of spaces and bodies, are crucial to identify, discuss, and critique—particularly in this moment—not only as large corporations are currently seeking to define and construct the metaverse, but also because policy makers are increasingly struggling with regulating such large corporate entities and their technological endeavours. The characteristics of the first two parts of this process, virtualization and datafication, are already quite clear in the present. The third part, in which the metaverse takes on an infrastructural role in society, is still emergent but its contours can nevertheless already be sketched and anticipated. In doing so, we hope to provide a starting point for opening up critical discussions about metaversal developments before the potential adverse effects of datafication become widespread and the foundations of this digital infrastructure have been constructed in the interests of corporations rather than in the interests of civil society.

Spaces and bodies, and particularly the reshaping and resignification of these phenomena through virtual environments and wearable technologies, lie at the centre of nearly all metaverse developments. What underpins such metaverse visions is the attempt at merging or blending of physical and digital spaces or practices. For instance, Pony Ma, head of the large Chinese technology corporation Tencent, has introduced the term 'immersive convergence' to describe "a new connection that integrates digital and physical forms, transcending time and space" (Tencent, 2022). Similarly, prominent investor and commentator, Matthew Ball, states in his widely-read book *The Metaverse and How It Will Revolutionize Everything* (2022, 97) that the metaverse "spans both the physical and virtual planes of existence". Indeed, virtual environments that create the impression of being 'somewhere' that is not physical reality are probably the most common interpretation of the metaverse, particularly in the form of virtual worlds, such as Second Life, Fortnite, and Roblox. But physical and spatial terms are not only employed in the context of platforms or worlds. For instance, Apple, during the release of its Vision Pro headset, strongly emphasized the notion of space, even going so far as describing the headset as "a revolutionary spatial computer that seamlessly blends digital content with the physical world", specifically by virtue of the "world's first spatial operating system" (Apple, 2023). As will be explained in more detail in Chap. 3, metaversal technologies and environments thus involve not just the creation of virtual spaces where people spend increasing amounts of time, but also the 'virtualization' or 'metaversification' of physical spaces through the development of digital twins as well as VR and AR technologies.

Furthermore, all metaverse visions emphasize the centrality of the body. For instance, Mark Zuckerberg described the metaverse as "an embodied internet where you're in the experience, not just looking at it" (Meta, 2021) during his announcement of Facebook's pivot towards the metaverse. In a later interview, Zuckerberg elaborates that the "embodied internet" involves not just "viewing content, you are in it" (Newton, 2021), suggesting that the boundaries between embodied experience and technology or media have almost fully disappeared. Perhaps unsurprising for a social media company, Meta heavily emphasizes social activities and states that the "defining quality" of the metaverse is that "you are right there with another person or in another place". Indeed, much metaversal development is centred upon emulating the appearance of face-to-face interaction, both in virtual environments or through headsets, thus simulating personal proximity and intimacy. Statements that emphasize the role of presence and immersion are important not just for understanding discourses around virtual spaces (e.g., being somewhere you are not physically able to be) but also for grasping the role of the body (being someone else). Importantly, immersion tends to be narrowly conceptualized as an embodied or sensorial state, thus leaving out an important dimension, namely how immersion is always also social and not just sensorial (Boellstorff, 2024). In contrast to previous discourses around embodiment during the emergence of the Web in the 1990s when 'disembodied' social interactions were emphasized (and sometimes thought to make identity categories such as gender redundant), the metaverse has given rise to new and important discussions about embodiment, both in terms of how identity can be expressed in metaverse spaces through virtual renditions such as avatars, but also how "feeling truly present with another person" (Meta, 2021) can be achieved, or at least simulated, through the use of metaversal technologies. Here, as with the notion of space, we see how embodiment is being reconfigured by its entry into virtual spaces and practices through the creation and use of virtual bodies or avatars, but also how the physical body and various physiological processes are being datafied and virtualized through the use of VR and AR headsets.

## 1.3  Metaversal Technologies, Platforms, and Environments

In addition to employing the 'metaverse' as an analytical term, despite the issues and uncertainties previously noted, it is also necessary to further define the term in order to provide it with more analytical traction. At this early and emergent stage, moreover, it is not prudent to depart from prominent definitions put forward by companies such as Meta or by investors and commentators such as Ball (2022), which have been partially or wholly adopted by some scholars and placed at the centre of popular discourse about the metaverse. This definition conceptualizes the metaverse as a predominantly virtual world that is somewhat separated from our physical reality, but we argue that the process of metaversification just as much applies to our physical spaces and bodies. Moreover, the 'metaverse' is surrounded as much by hype as by anti-hype, and therefore subject to cycles of both extreme enthusiasm and criticism in which the presentation of nuanced analyses is overshadowed by the taking of polemical positions. Sensational promotion and rampant speculation not only surround the term itself but also many of its purported core aspects, constituent parts, and key commodities, such as virtual worlds, gaming platforms, blockchains, cryptocurrencies, and non-fungible tokens (NFTs), as well as wearable devices, such as Meta's Quest and Apple's recently released Vision Pro, which have all been described as either paradigm-shifting technologies or boondoggles. Cryptocurrencies and tokenized assets such as NFTs, in particular, appear to be prone to strong fluctuations in both popularity and value, indicating that their meaning and function are far from stabilized and thus their implications far from certain. Furthermore, as already noted, even a brief glance at the ways in which various actors describe 'the metaverse' shows that the word can mean many and sometimes contrasting things. There is, for example, little agreement on whether to capitalize or pluralize the term, whether to consider the metaverse as a (de)centralized platform or instead as a mobile or wearable technology akin to the smartphone, or whether the metaverse is already here, going to arrive soon, or impossible to achieve even in the not-so-near future. For all these reasons, we propose an admittedly very broad and capacious definition that seeks to capture the diverse range of technologies and platforms that are involved in the emergent process of virtualization, datafication, and infrastructuralization of both spaces and bodies. But first it is necessary to briefly examine and outline how technology companies, commentators, and researchers broadly as well as diversely conceptualize the metaverse.

Several general features of the metaverse and distinctions between disparate approaches can be identified from existing communications from technology companies and industry reports. One of the main (and older) narratives that informs these definitions is one that portrays the metaverse as a long-awaited culmination of developments in virtual reality (VR) and augmented reality (AR) technology, which are both commonly associated with the emergence of video games since the 1980s. In VR, one's physical environment is replaced almost entirely with a computer-generated representation. Instead in AR, one's visual field of the physical environment is overlaid with digital elements or layers, which can convey, for example, information or directions. A crucial distinction between VR and AR is, therefore, that the former tends to fully envelop or immerse those who seek to enter such virtual environments or 'worlds', whereas the latter retains a considerable level of direct perception and experience of the physical environment (Bolter et al., 2021). Both VR and AR are sometimes, rather confusingly, also referred to as extended reality (XR) or mixed reality (MR). XR appears most commonly used as an umbrella term to indicate both VR and AR. MR, however, tends to indicate a state that lies somewhere between the fully computer-simulated experience of VR and the relatively low level of simulation in AR. More precisely, MR is used to describe a situation in which the digital elements that are overlaid on one's field of perception can be interacted with or manipulated through an interface (whether visual, haptic, or other) rather than merely noted as items of information. Although a debatable distinction, the suggestion is that MR allows for more interactivity whereas AR implies a more passive and merely informational rather than interactive dynamic.

More broadly speaking, VR and AR technologies—despite also being used for and developed in military, medical, and educational contexts (Carter & Egliston, 2024) —tend to be most closely associated with video games and gaming platforms. Perhaps the most famous example of a gaming application that can be described as either AR or MR is the smartphone game Pokémon Go, which was released in 2016 to great popular acclaim, and continues to attract a large number of monthly players. In terms of VR, the most popular gaming titles are Beat Saber and Half Life: Alyx, which are, respectively, a rhythm-based game involving the slicing of blocks to the beat of a song, and a classic first-person shooter set in an alien dystopia. However, although popular VR and AR games as well as VR-exclusive platforms do exist, and several major metaverse companies such as Meta are heavily invested in developing VR technologies, the most common means of accessing most gaming platforms, such as Roblox and Fortnite, still remains relatively conventional, namely through personal computers, game consoles, and smartphones. In short, it is probably unhelpful to conceptualize the metaverse as defined by access through VR interfaces (Boellstorff, 2024). Consequently, it is still up for debate whether the metaverse is more an extension of the decades-old push towards developing VR or rather an entirely new development (Evans, 2019; Harley, 2024). This perspective, in other words, presents the metaverse generally as a type of media or game that is generally consumed or played, but not necessarily exclusively, through specialized equipment such as head-mounted displays, smart glasses, and haptic interfaces. As such, the metaverse is more aligned with video games and massively multiplayer online games (MMOGs), which can be accessed and played through conventional devices on separate and quite distinct platforms. Because of this plurality, researchers tend to refer to these developments in their plural form, namely as platforms, metaverses, or even 'microverses' rather than as a single entity or -*verse* (Evans et al., 2022).

More recently and significantly less modestly, the metaverse is also frequently represented—particularly in communications from technology companies and consultancies—as the continuation of or next step in the evolution of the internet, and particularly as the future of mobile internet and smartphone technology. From this perspective, the emphasis is less on gaming and more on a combination of social media applications and wearable devices that are or will be used in various professional and industrial as well as leisure settings (McKinsey, 2022), frequently with a strong emphasis on AI or so-called Web3 technologies, such as blockchains, cryptocurrencies, and NFTs. As noted previously, the metaverse is presented by technology companies as the 'embodied internet' or a 3D immersive version of the internet that can be entered and occupied rather than merely accessed (e.g. Meta, 2021). Whereas Meta and Zuckerberg have closely associated themselves with the term 'Metaverse' (capitalized and including the definite article *the*), most notably through the company's name change from Facebook to Meta in 2021, other companies, such as Google and Apple, prefer to use the terms 'ambient computing' or 'spatial computing', which imply a similar degree of both ubiquity and persistency, but not necessarily in an entirely new or virtual space but instead integrated in the physical environment. Here, the metaverse is conceptualized less as a form of media and more as a device and/or platform in, through, or upon which other digital technologies such as apps can be developed or used.

In its most extreme or radical form—most famously espoused by Meta— 'The Metaverse' is presented as a single interconnected world in which many if not almost all professional, social, and leisure activities can supposedly take place. As such, the metaverse is conceptualized not merely as a space that is occasionally visited or used to, say, play a game, but as an infrastructure that is permanently and consistently present and all-encompassing in everyday life as well as underpinning a much wider and more important range of uses compared to current media, technologies, or platforms. In its grandest and perhaps most outlandish form—as a single virtual world encompassing almost all if not all digital activities—this version of the metaverse is deemed by many to be technically feasible but not necessarily achievable within the very near future. Advances in computing (particularly graphics and processions power), connectivity (high bandwidth and low latency), battery power, miniaturization, in addition to unifying, standardized software protocols to ensure interoperability, are all necessary improvements before this grand vision can be achieved. And this is not to mention the new complexities, vulnerabilities, and concerns (e.g. cybercrime, mis- and disinformation, harassment, sustainability (both social and environmental) that will arise once such technical achievements make widespread virtuality possible (Lee et al., 2024).

This all-encompassing vision is still only a future projection, which, even if more hype than reality and even if only partially achieved, will nonetheless have a major impact on society, ranging from the emergence of new types of social interaction to enabling new forms of entrepreneurship and surveillance. For now, however, the contemporary landscape is characterized by a plurality of platforms and devices, some of which seem relatively stabilized and interconnected whereas others appear relatively experimental and insular. Moreover, these currently existing platforms and devices are already indicative of the datafication efforts that underpin their construction as well as necessary steps towards the construction of the grand-vision metaverse. Current platforms that are frequently categorized as metaversal include, among others: Second Life, Horizon Worlds, VRChat (socializing and work); Roblox, Minecraft, Spatial, Fortnite (gaming and game creation); Decentraland, The Sandbox, Axie Infinity (investment in virtual land and cryptocurrencies in combination with gamified 'experiences'); Nvidia Omniverse, Microsoft Mesh (digital twins for enterprises as well as platforms for professional collaboration). However, it must be noted that the distinctions between, say, 'gaming' and 'work' are somewhat arbitrary as platforms such as Roblox, for example, are seeing the emergence of professionalized labour (e.g., 'playbour') and business models (e.g., monetization of user-generated content) that are similar to the role of influencers and creators on social media platforms such as Instagram, YouTube, and TikTok (Foxman, 2022; Nieborg & Poell, 2018). All these metaverse platforms are characterized by large differences in terms of their underpinning technologies (VR, AR, or 'conventional' 2D worlds), design and organization (centralized or decentralized), operations and business models (advertising, subscriptions, user-generated content), internal economy ('real' fiat currencies, on-platform or in-game currencies, or cryptocurrencies and assetization through blockchains) modes of access (purchase, subscription, free-to-play), practices (gaming, content creation, socializing, professional labour), and popularity (ranging from tens if not hundreds of millions users to a few thousand or even fewer). In other words, there is a great diversity in terms of how these platforms operate, and it is uncertain exactly which platforms will succeed at hosting what kinds of activities, or which technologies will be utilized for which professional or leisure purposes.

Whether the 'metaverse' is best conceptualized as a wearable device, game, media, platform, or as a world, or either as some new combination of all or as something else entirely, will inevitably be determined in the near future as platforms rise and fall and certain technologies achieve commercial success over others. In the business world, there is a fervent hope for a killer app or device that can finally open the 'metaverse market' to mass consumption. It is important to note, however, that the narratives currently swirling around the metaverse not only closely mirror the discourses surrounding the emergence of the commercial internet in the 1990s and the introduction of mobile computing in the 2000s (as well as the colonialist metaphors), but also actively seek to situate the metaverse within this historical evolution in order to frame and justify its technological development and cultural relevance as well as in the hope attracting further investment. In short, the metaverse is framed as a linear and logical next step that will inevitably follow from as well as supersede internet and smartphone technology. Consequently, it is tacitly assumed, the metaverse will match the role and impact that these preceding technologies have had on society. At the moment, therefore, the label 'metaverse' clearly plays an important anticipatory and future-making role for technology companies in their discursive constructions of its economic and socio-cultural relevance. Although this aspect of metaverse discourse as a process of 'worldmaking' (Haraway, 2016) is crucial to address, it is also important to note the backward-looking and revisionary uses of the term as it is applied to previous virtual environments (e.g. Second Life) as well as current gaming platforms (e.g. Minecraft, Roblox, Fortnite), which are being retrospectively redefined by commentators and researchers as metaversal (Boellstorff, 2022; Perry, 2022).

The emergent scholarly literature in the social sciences and humanities on the metaverse is similarly characterized by a plurality of definitions, inconsistencies of usage, and disagreements on how to conceptualize the metaverse, how this might be accessed or used, or what the ancillary or underpinning technologies might be. Many scholars, moreover, prefer to focus on more conventional and established terms such as Virtual Reality, Augmented Reality, Mixed Reality, and Extended Reality (Bolter et al., 2021; Carter & Egliston, 2024; Evans, 2019; Harley, 2024; Messeri, 2024), and to consider these under the umbrella of the 'metaverse' (if this is explicitly mentioned at all). Perhaps unsurprisingly, much research in the Anglophone literature tends to centre on US-based companies and particularly on the operations of Meta and its research unit Reality Labs, or on large gaming platforms such as Roblox and Fortnite, while ethnographic studies on the development and use of virtual environments or technologies remain few and far between (notable exceptions being Boellstorff, 2008; Evans, 2019; Malaby, 2009; Messeri, 2024). Consequently, global development of the metaverse in places outside of the Euro-American context tends to be underacknowledged (Girginova, 2024), although there are notable exceptions (e.g., Negro & Savina, 2024; Roquet, 2023). Furthermore, there are several definitional disagreements and tensions that exist in the current critical literature on the metaverse. For example, some prominent researchers do not consider VR to be a necessary means for accessing the metaverse (Boellstorff, 2024). Similarly, massiveness, interoperability, and blockchain technology, which some prominent commentators and stakeholders have claimed to be fundamental features of metaverse 'worlds' (e.g. Ball, 2022; Lee et al., 2024), are not necessarily seen as such by others who consider it unclear why intercommunication between massive-scale 'worlds' or carrying out commercial activity through distributed ledgers are necessarily a prerequisite for metaverses to emerge (Boellstorff, 2024).

Nevertheless, it is notable that several academic studies appear to take at face value the definition of the metaverse as put forward by prominent tech companies, such as Meta, or venture capital firms, such as Andreessen Horowitz. Particularly influential in this regard has been the definition put forward by entrepreneur and venture capitalist (as well as early investor in Facebook) Matthew Ball in his 2022 book *The Metaverse: And How It Will Revolutionize Everything*:

> The Metaverse is a massively scaled and interoperable network of real-time rendered 3D virtual worlds that can be experienced synchronously and persistently by an effectively unlimited number of users with an individual sense of presence, and with continuity of data, such as identity, history, entitlements, objects, communications, and payments (Ball, 2022, 29)

Ball's definition is remarkably close to those employed by some academic researchers and commentators (e.g. Au, 2023; McStay, 2023; Mosco, 2023) and even included in meta-analyses of the scholarly literature (e.g. Ritterbusch & Teichmann, 2023; Weinberger, 2022). Perhaps more problematic is that the corporate and venture-capitalist framing and positioning of the metaverse as both an inevitability as well as necessity is similarly taken for granted by some researchers and commentators. Although it is absolutely true that the amount of resources dedicated to metaverse projects warrants academic scrutiny, that the metaverse visions of big technology corporations needs to be taken seriously even if these are only partially achieved, and that corporate narratives certainly are important subjects of inquiry, it makes little sense for researchers to ostensibly reproduce the suggestion that the metaverse will revolutionize, disrupt, or transform 'everything' in the way that corporate actors insist on (Natale et al., 2019). Instead, we believe it more fruitful to examine how metaverse platforms and technologies—in the here and now—are being applied in the pursuit of the interests and goals of a small but very powerful number of companies.

A further consequence of hewing too closely to corporate definitions of the metaverse as predominantly a virtual space is that this threatens to draw attention away from how companies can and already are collecting information about physical spaces and bodies, and not just through headsets but also other mobile or wearable devices. Companies, such as Google and Meta in particular, are also already positioning themselves as infrastructural actors, as can be witnessed from their involvement in the laying of intercontinental submarine cables that form the backbone of internet communication. In other words, the question of datafication in and through metaverse technologies is rarely addressed, and even when addressed, this is only done partially because the focus tends to be on either spaces or bodies separately. Instead, we believe it is necessary to highlight the continuities between the virtualization and datafication practices that have emerged during the so-called era of Big Data—often initiated and exacerbated by current 'Big Tech' companies that are heavily investing in the metaverse—and to explore how the current emergence and deployment of metaversal technologies can be understood as an intensification of current datafication practices and a preliminary step in the move towards infrastructuralization.

## 1.4   Metaversification of Spaces and Bodies

Rather than further speculating on the potential importance of interoperability, synchronicity, persistence and scale of metaverses, or whether VR interfaces are conceptually distinct from virtual places or environments, we first focus on the current and predominant technological components of the metaverse and particularly their potential in the present for deepening and intensifying the datafication of our bodies, behaviours, and spaces. The term 'metaversal technologies' can usefully be employed to describe the spectrum of devices and processes that enable the collection of 'metaversal data', that is to say, a more extensive type of data that is collected both on the bodies of people as well as the environments in and through which they work, live, and interact. This process of datafication is, as already observed, an initial step towards infrastructuralization.

A first set of metaversal technologies encompasses the emerging range of wearable (generally head-mounted) and sensory (i.e. spatially-aware as well as body-tracking) computing devices. Such technologies are referred, as already noted, under a range of different names, such as virtual reality (VR), augmented reality (AR), mixed reality (MR), and extended reality (XR), which all refer to the mutual incorporation of physical and digital dimensions of the environment as well as the 'user' or 'wearer' of these devices. These devices all have in common the extension of virtual layers onto the perceptual field for the purpose of enhancing various activities, which can range from gaming and other forms of audio-visual entertainment to attending work meetings and carrying out professional activities. Crucially, these devices are capable of tracking not only the behaviours and actions of wearers as well as a range of their biometric characteristics and processes, but—equally importantly—also the features and characteristics of the environments in which wearers are situated and move as well as their interactions with other beings and things in this space.

Mixed-reality technologies necessarily have to capture and process data about the user's body in order to make its movements legible to the computing system as well as to generate and project images or virtual layers that are both perceptible and manipulable by the user (Egliston & Carter, 2021; Golding, 2019). The types of physiological data that can potentially be captured through mixed-reality technologies range from information about one's face, eyes, gaze, and voice to the characteristic movements of one's limbs and gait. Current mixed-reality devices, including the most popular commercial devices currently on the market, such as the Quest headset sold by Meta and the Vive headset produced by HTC, are capable of collecting data from users that count as biometric (Egliston & Carter, 2023).

In addition to the integration of biometric and sensory data, which have long been difficult to capture and combine with current wearable devices, mixed-reality technologies are also increasingly capable of furnishing another type of data, namely about the spaces and environments in which such technologies are worn. To track and make legible the body of a user, mixed-reality technologies also need to sense the environment in which this body (as well as the device itself) are situated. In particular, fully-mobile headsets, i.e., not tethered to or reliant on another device, employ odometric sensors, computer vision, and algorithms or AI to map their surrounding environments and to track the position and movement of both users and their devices (Egliston & Carter, 2023, 2024; Saker & Frith, 2020). This allows for the extraction and collection of data about, for example, public infrastructures and spaces, workplaces, and 'home life' data (Barassi, 2020) as well as the composition of these spaces and the objects and people found therein. In other words, creating the sense of immersion and presence that metaverse technologies aim to achieve not only relies on different kinds of data but also allows for the collection of new and different kinds of embodied, spatial, and environmental data. In an important sense, such devices can be understood as a means for pushing the metaverse and the logic of virtualization into the existing world.

A second category of metaverse technologies encompasses the efforts being made to create and operate persistent and immersive virtual environments or 'worlds' in which a range of professional, social, and leisure activities are taking place or purportedly will do so. This is probably the most common popular interpretation of the term 'metaverse', in part because of comparisons that have been drawn with previous virtual worlds, such as Second Life, or gaming environments, such as the wildly popular game Fortnite. Another reason for the predominant interpretation of 'metaverses' as virtual 'worlds' lies in the widely reported name change by Facebook to Meta and its subsequent presentation of the Horizon Worlds platform. However, current communications from metaverse companies such as Meta appears to be shifting towards a definition of 'metaverse' that is less reliant on virtual worlds and more on headset-mediated forms of Mixed Reality. Meta's Quest headsets, for example, are no longer referred to as VR but instead as MR. In comparison to data collected through metaversal devices such as headsets, the current literature remains somewhat sparse and relatively ambiguous about the applications and implications of data that can be collected through gaming platforms and virtual worlds. Nevertheless, what is clear is that data can be collected on in-game actions, behaviours, and purchases as well as interactions with other people in virtual worlds and the environment, or personally identifiable information (e.g. name, location, age, gender), and device details (e.g. hardware specifications and software versions) (Seif El Nasr et al., 2013). In contrast to the wearable devices that push the metaverse into our physical environment, virtual worlds can be understood as a means of pulling existing practices, services, processes into the metaverse.

Metaverse data is not an entirely new phenomenon that radically breaks with the era of big data but an expansion of this process of datafication into important new domains of human activity. Such human activities can take place in newly constructed virtual environments or 'worlds', but this is certainly neither the only nor the most important means through which metaverse datafication is carried out. Indeed, the availability and commercial success of mixed-reality headsets, which are gradually becoming a mass-market phenomenon, will both deepen and expand the data that can be collected on spaces and bodies. The metaverse, at least at this point in time, is as much about creating virtual worlds from scratch as it is an attempt at what can be called 'pandatafication'—the process through which the entire world can be quantified, measured, and reduced to what can be read, navigated, and ultimately managed by automated systems such as AI (Hesselbein et al., 2024a). Because of its emergent character, we opt for an intentionally broad definition of metaverse technologies to capture the potential full range of datafication practices and their implications for subsequent infrastructuralization. Research should not only focus on virtual environments or worlds, but also on the broad range of ancillary devices, such as VR headsets, AR glasses, and haptic gloves, that can be used either to access such virtual worlds or to 'virtualize' us and our environments. What is more, it is the coupling of such granular 'physical' data and 'virtual' data, collected through different devices or platforms and involving different practices, that is thought to be a particularly powerful combination that will provide a much more intimate insight into our lives to those who can collect, aggregate, and exploit such data (Evans et al., 2022). Above all, we need to keep in mind that the metaverse is not just a technology or platform, but also a process of metaversification that seeks to draw in existing activities into the ambit of 'tech' companies as well as to expand their sphere of influence outwards to places and practices that have not yet been captured and monetized.

## 1.5   Background and Approach

This book is both exploratory and anticipatory in nature. Drawing on our disciplinary backgrounds in Science and Technology Studies (STS) and Media and Communication Studies, our first objective is to critically examine and analyse the current activities and developments as well as future visions of technology companies seeking to construct the metaverse. Our focus is primarily on the virtualization and datafication of spaces and bodies, and how this process can prepare the ground for subsequent infrastructuralization. The characteristics of the first part of this process are already quite clear in the present. The second part, in which the metaverse is likely to take on an infrastructural role in society, is still emergent, but its contours can nevertheless already be sketched and anticipated. In addition to highlighting the datafication of society through metaversal technologies, the second objective of this book is therefore to provide an early and anticipatory discussion of the infrastructural implications of the metaverse. Although predominantly focusing on recent, current, and near-future developments by technology companies, we believe it also necessary to retrace the historical origins of the metaverse from a cultural perspective, thus underscoring the role of fiction, media, and technology narratives and how these can provide a corrective and alternative imagination to the visions emanating from large technology corporations. Last, it must be noted that we are mainly focusing on two important aspects of infrastructuralization, namely the capturing and reshaping of spaces and bodies. Other aspects, such as the ways in which the metaverse can be understood as a means of seizing more of our time and attention, or as a potential means for absorbing financial services, are important avenues for future research. In this book, moreover, we will also pay less attention to the role of data infrastructures, such as submarine and terrestrial cables, wireless networks, data centres, servers, and energy infrastructures, which are all essential to the future functioning of 'the metaverse'. These are equally important infrastructures of reality that require further reflection yet that go beyond the scope of this book.

Quite a bit of our analytical focus will be on Meta, because this company is the main driver behind much of the metaverse hype and because its vision is the most likely one to become dominant in the future. This does not mean that Meta's version of the metaverse will necessarily become the dominant one, but we do need to take its role in the development of the metaverse seriously since it is one of the most well-resourced actors and its vision of the metaverse is perhaps more elaborated and articulated than that of others. Meta's version of the metaverse is simultaneously old and new (Bell, 2022). It harkens back to old hopes and hypes around VR, which in many ways already exist in commercially available 'microverses' (Evans et al., 2022), but with the ambition of connecting such separate worlds into a single, full-blown, interconnected, and interoperable Metaverse (with a capitalized M). Part of the 'oldness' of Meta's version of the metaverse can be traced back to sci-fi literature, and the novel *Snow Crash* in particular (Au, 2023), which is why we dedicate a separate chapter to exploring such narratives and their impact on both imaginaries of the metaverse as well as its construction. Furthermore, as noted at the very start, other technology companies have different visions of what the metaverse is, will, or should be, but they all nonetheless converge on the idea that the digital will subsume the physical in an inexorable manner. While focusing on the datafication of spaces and bodies by 'Big Tech', the visions of other companies working towards the metaverse, some of which have a relatively low public profile, will also be included in our analysis.

Our book takes a critical approach to the metaverse, as least as long as it is being built in the image and according to the specific goals of large technology companies, particularly because we know that many of these companies have spent the past decades securing their monopolistic positions as 'Big Tech' while hoovering up as much data as they could possibly grab. As we now know, this data has been crucial for the developments of AI systems, which are now being touted as yet another means of disrupting and revolutionizing 'everything'. It is therefore an entirely open question if the metaverse is worth creating at all, particularly if it built and largely controlled by such companies. As we know since the Covid-19 pandemic in particular, the technologies and services of technology companies can allow a great deal of personal and professional life to more or less continue, but we also know that the benefits of remote working and the reliance on such infrastructure has come at a cost. Moreover, the metaverse is also a term at the centre of growing geopolitical tensions between the US, the EU, and China, which are each seeking to gain strategic advantage in the so-called AI arms race as well as the development of digital infrastructures. Big technology projects, and especially infrastructural ones, therefore give rise to new forms of competition and vulnerability.

Although our starting point is critical, we certainly do not want to deny that the metaverse can potentially offer great benefits and even pleasure and fulfilment. Genuine communities can be created and sustained in virtual worlds (Boellstorff, 2008), and otherwise marginalized people can make new connections, ones that might have been impossible in more conventional places. Telepresence can provide a much-needed connection between those who are far away or those who struggle to leave their homes. What is more, VR technology is proving to be a useful therapeutic tool as well as an impressively fun leisure device. Others have sought to employ virtual technologies for the preservation of cultural heritage, the safeguarding of indigenous languages and traditions, and for the exploration of new creative and highly imaginative futures. Indeed, there is a real possibility that the metaverse can be used for fostering new types of civil society and for working towards a common good that is truly for everyone rather than just for those who have a stake in technology companies.

This book is structured as follows. In Chap. 2, we discuss how the metaverse is only the latest in a long line of utopian and dystopian visions of a digital future in which being simultaneously 'here and elsewhere' will, it is claimed, revolutionize social interaction, the economy, and society at large. Previous technological projects and the narratives surrounding them take on very different meanings in different contexts and result in different social expectations or even political goals. This chapter lays the groundwork for the upcoming two chapters by tracing the main narratives about virtual worlds and their historical origins—including their escapist visions of disembodiment and spatiality—back to scientific and fiction literature, sci-fi movies, and corporate and political documents.

In Chap. 3, we examine how metaversal technologies are involved in the reconfiguration of spatiality, particularly on the level of virtualization and datafication. We first explore the various ways in which leading companies conceptualize space in relation to their own business interests, generally by setting up virtual worlds into which they seek to draw in various social activities while simultaneously giving rise to new forms of labour and opportunities for monetization and exploitation. Next, we discuss how virtualization processes, such as the employment of game engines and the creation of digital twins for capturing and controlling a wide variety of real-world structures and objects, are positioning 'tech' companies as a central infrastructure across various industries. Last, we discuss how wearable VR and AR devices are used to capture physical spaces through virtualization and datafication, thus reconfiguring spatiality, and what the implications of this are for the development of emergent AI systems.

Chapter 4 focuses on the topic of corporeality and how metaversal technologies can reconfigure our sense of embodiment, particularly through the reproduction of not just our physical bodies but also their virtual renditions. We start by examining how 'tech' companies conceptualize embodiment and its role in professional as well as private life as well as its sensory, experiential, and cognitive capacities. Next, we focus on how virtual worlds are positioned as new spaces for the expression of identity through the consumption of virtual fashion and apparel while simultaneously commodifying existing forms of social interaction, economic production, and self-presentation. Then we turn to examining how metaversal technologies, such as VR and AR devices, can not only to collect new and more extensive forms of data about bodies, but also to extend as well as narrowly redefine our corporeal capacities and individual identities. Last, we reflect on some of the connections and implications of the virtualization and datafication of bodies, particularly for the development of AI systems that are becoming increasingly present in virtual as well as physical environments.

In the Conclusion, we underline the main thrust of the book's argument and discuss its implications in terms of spatiality and corporeality. Specifically, we underscore our reformulation of the term 'metaverse' as a process of metaversification in which existing activities and practices are drawn into virtual environments while the logics of virtuality are simultaneously pushed outwards into our environments and bodies. We subsequently also reflect on the specific nature of the various companies involved in this process of metaversification and the importance of developing alternative narratives that might counter or even subvert hegemonic visions of the metaverse. Next, we briefly discuss how the reality of infrastructures in terms of their material requirements should be carefully considered, especially before committing considerable resources in this direction. Finally, we raise a number of suggestions for critiquing as well as potentially countering corporate and privatized visions of the metaverse with a specific focus on the ways in which it should be considered a communal and public good.

# References

Apple. (2023, June 5). *Press release. Introducing Apple Vision Pro: Apple's first spatial computer.* Retrieved March 15, 2025, from https://www.apple.com/newsroom/2023/06/introducing-apple-vision-pro/

Au, W. J. (2023). *Making a metaverse that matters: From snow crash & second life to a virtual world worth fighting for.* Wiley.

Ball, M. (2022). *The metaverse: And how it will revolutionize everything.* Liveright Publishing Corp.

Barassi, V. (2020). *Child data citizen: How tech companies are profiling us from before birth.* MIT Press.

Barbrook, R., & Cameron, A. (1996). The Californian ideology. *Science as Culture, 6*(1), 44–72. https://doi.org/10.1080/09505439609526455

Beer, D. (2024). Extensive culture: Expressions of endlessness in the metaverse and the limits of data accumulation. *Information, Communication & Society, 28*(5), 926–941. https://doi.org/10.1080/1369118X.2024.2413114

Bell, G. (2022, February 8). The metaverse is a new word for an old idea. *MIT Technology Review.* https://www.technologyreview.com/2022/02/08/1044732/metaverse-history-snow-crash/

Boellstorff, T. (2008). *Coming of age in second life: An anthropologist explores the virtually human.* Princeton University Press.

Boellstorff, T. (2022, June 15). How we describe the metaverse makes a difference—Today's words could shape tomorrow's reality and who benefits from it. *The Conversation.* https://theconversation.com/how-we-describe-the-metaverse-makes-a-difference-todays-words-could-shape-tomorrows-reality-and-who-benefits-from-it-182819

Boellstorff, T. (2024). Toward anthropologies of the metaverse. *American Ethnologist, 51*(1), 47–56. https://doi.org/10.1111/amet.13228

Bolter, J. D., Engberg, M., & MacIntyre, B. (2021). *Reality media: Augmented and virtual reality.* MIT Press.

Boyd, D., & Crawford, K. (2012). Critical questions for Big Data. *Information, Communication & Society, 15*(5), 662–679. https://doi.org/10.1080/1369118X.2012.678878

Carter, M., & Egliston, B. (2024). *Fantasies of virtual reality: Untangling fiction, fact, and threat.* MIT Press.

Couldry, N., & Meijas, U. A. (2019). *The costs of connection: How data is colonizing human life and appropriating it for capitalism.* Stanford University Press.

Egliston, B., & Carter, M. (2021). Critical questions for Facebook's virtual reality: Data, power and the metaverse. *Internet Policy Review, 10*(4). https://policyreview.info/articles/analysis/critical-questions-facebooks-virtual-reality-data-power-and-metaverse

Egliston, B., & Carter, M. (2023). Examining visions of surveillance in Oculus' data and privacy policies, 2014–2020. *Media International Australia, 188*(1), 52–66. https://doi.org/10.1177/1329878X211041670

Egliston, B., & Carter, M. (2024). 'The metaverse and how we'll build it': The political economy of Meta's Reality Labs. *New Media & Society, 26*(8), 4336–4360. https://doi.org/10.1177/14614448221119785

Evans, L. (2019). *The re-emergence of virtual reality.* Routledge.

Evans, L., Frith, J., & Saker, M. (2022). *From microverse to metaverse: Modelling the future through today's virtual worlds.* Emerald Group Publishing.

Fourcade, M., & Healy, K. (2017). Seeing like a market. *Socio-Economic Review, 15*(1), 9–29. https://doi.org/10.1093/ser/mww033

Foxman, M. (2022). Gaming the system: Playbour, production, promotion, and the metaverse. *Baltic Screen Media Review, 10*(2), 224–233. https://doi.org/10.2478/bsmr-2022-0017

Girginova, K. (2024). Global visions for a metaverse. *International Journal of Cultural Studies, 28*(1), 300–306. https://doi.org/10.1177/13678779231224799

Golding, D. (2019). Far from paradise: The body, the apparatus and the image of contemporary virtual reality. *Convergence, 25*(2), 340–353. https://doi.org/10.1177/1354856517738171

Haraway, D. (2016). *Staying with the trouble: Making kin in the Chthulucene.* Duke University Press.

Hardaker, S. (2025). From bytes to bricks: Advocating for a turn toward platform-led infrastructuralization in economic geography. *Progress in Economic Geography, 3*(1), Article 100038. https://doi.org/10.1016/j.peg.2025.100038

Harley, D. (2024). "This would be sweet in VR": On the discursive newness of virtual reality. *New Media & Society, 26*(4), 2151–2167. https://doi.org/10.1177/14614448221084655

Hendrikse, R., Adriaans, I., Klinge, T. J., & Fernandez, R. (2022). The big techification of everything. *Science as Culture, 31*(1), 59–71. https://doi.org/10.1080/09505431.2021.1984423

Hesselbein, C., Bory, P., & Canali, S. (2024a). Metaverse datafication: Technologies, definitions, and futures. *Information, Communication & Society, 28*(5), 763–777. https://doi.org/10.1080/1369118X.2024.2443082

Hesselbein, C., Bory, P., & Canali, S. (2024b). Six provocations for metaverse datafication: An emergent cultural, technological, and scholarly phenomenon. *Information, Communication & Society, 28*(5), 778–796. https://doi.org/10.1080/1369118X.2024.2433548

Hwang, T. (2020). *Subprime attention crisis: Advertising and the time bomb at the heart of the internet.* FSG Originals x Logic.

Kerssens, N., & van Dijck, J. (2023). Transgressing local, national, global spheres: The black-boxed dynamics of platformization and infrastructuralization of primary education. *Information, Communication & Society, 27*(15), 2600–2616. https://doi.org/10.1080/1369118X.2023.2257293

Lee, L. H., Braud, T., Zhou, P. Y., Wang, L., Xu, D., Lin, Z., Kumar, A., Bermejo, C., & Hui, P. (2024). All one needs to know about metaverse: A complete survey on technological singularity, virtual ecosystem, and research agenda. *Foundations and Trends in Human-Computer Interaction, 18*(2–3), 100–337. https://doi.org/10.1561/1100000095

Malaby, T. M. (2009). *Making virtual worlds: Linden lab and second life.* Cornell University Press.

McKinsey. (2022). *Value creation in the metaverse.* Retrieved March 15, 2025, from https://www.mckinsey.com/capabilities/growth-marketing-and-sales/our-insights/value-creation-in-the-metaverse

McStay, A. (2023). The metaverse: Surveillant physics, virtual realist governance, and the missing commons. *Philosophy & Technology, 36*(13), 1–26. https://doi.org/10.1007/s13347-023-00613-y

Messeri, L. (2024). *In the land of the unreal: Virtual and other realities in Los Angeles.* Duke University Press.

Meta. (2021) *Founder's letter, 2021.* Retrieved March 15, 2025, from https://about.fb.com/news/2021/10/founders-letter

Microsoft. (2021). *Ignite 2021: Satya Nadella keynote transcript.* Retrieved March 15, 2025, from https://news.microsoft.com/wp-content/uploads/prod/2021/11/Microsoft-Ignite-2021-Satya-Nadella.pdf

Mosco, V. (2023). Into the metaverse: Technical challenges, social problems, utopian visions, and policy principles. *Javnost-the Public, 30*(2), 1–13. https://doi.org/10.1080/13183222.2023.2200688

Natale, S., Bory, P., & Balbi, G. (2019). The rise of corporational determinism: Digital media corporations and narratives of media change. *Critical Studies in Media Communication, 36*(4), 323–338. https://doi.org/10.1080/15295036.2019.1632469

Negro, G., & Savina, T. (2024). Yuanyuzhou 元宇宙: yesterday, today, tomorrow. Historical roots, current visions, and future dynamics of real-world integration in the Chinese governmental narrative on the Metaverse. *Information, Communication & Society, 28*(5), 890–909. https://doi.org/10.1080/1369118X.2024.2442390

Newton, C. (2021, July 22). Mark Zuckerberg is betting Facebook's future on the metaverse. *The Verge.* https://www.theverge.com/22588022/mark-zuckerberg-facebook-ceo-metaverse-interview

Nieborg, D. B., & Poell, T. (2018). The platformization of cultural production: Theorizing the contingent cultural commodity. *New Media & Society, 20*(11), 4275–4292. https://doi.org/10.1177/1461444818769694

Perry, A. (2022, September 2). The "real" metaverse already exists and it's called "Fortnite". *Mashable.* https://mashable.com/article/fortnite-is-the-real-metaverse

Plantin, J.-C., Lagoze, C., Edwards, P. N., & Sandvig, C. (2018). Infrastructure studies meet platform studies in the age of Google and Facebook. *New Media & Society, 20*(1), 293–310. https://doi.org/10.1177/1461444816661553

Ritterbusch, G. D., & Teichmann, M. R. (2023). Defining the metaverse: A systematic literature review. *IEEE Access, 11,* 12368–12377. https://doi.org/10.1109/access.2023.3241809

Roquet, P. (2023). Japan's retreat to the metaverse. *Media, Culture & Society, 45*(7), 1501–1510. https://doi.org/10.1177/01634437231182001

Ruckenstein, M., & Schüll, N. D. (2017). The datafication of health. *Annual Review of Anthropology, 46*(1), 261–278. https://doi.org/10.1146/annurev-anthro-102116-041244

Sadowski, J. (2019). When data is capital: Datafication, accumulation, and extraction. *Big Data & Society, 6*(1), 1–12. https://doi.org/10.1177/2053951718820549

Saker, M., & Frith, J. (2020). Coextensive space: Virtual reality and the developing relationship between the body, the digital and physical space. *Media, Culture & Society, 42*(7–8), 1427–1442. https://doi.org/10.1177/0163443720932498

Seif El-Nasr, M., Drachen, A., & Canossa, A. (2013). *Game analytics: Maximizing the value of player data.* Springer.

Shapiro, E. (2021, April 18). Nvidia CEO Jensen Huang talks the powers of automation. *Time.* https://time.com/5955412/artificial-intelligence-nvidia-jensen-huang/

Tencent (2022, September 29). *Tencent introduces 'immersive convergence' to drive connections between digital and real worlds.* Retrieved March 15, 2025, from https://www.tencent.com/en-us/articles/2201445.html

Thatcher, J., O'Sullivan, D., & Mahmoudi, D. (2016). Data colonialism through accumulation by dispossession: New metaphors for daily data. *Environment and Planning D: Society and Space, 34*(6), 990–1006. https://doi.org/10.1177/0263775816633195

Van der Vlist, F., Helmond, A., & Ferrari, F. (2024). Big AI: Cloud infrastructure dependence and the industrialisation of artificial intelligence. *Big Data & Society, 11*(1), 1–16. https://doi.org/10.1177/20539517241232630

Van Dijck, J. (2014). Datafication, dataism and dataveillance: Big Data between scientific paradigm and ideology. *Surveillance & Society, 12*(2), 197–208. https://doi.org/10.24908/ss.v12i2.4776

Weinberger, M. (2022). What is Metaverse? A definition based on qualitative meta-synthesis. *Future Internet, 14*(11), 310. https://doi.org/10.3390/fi14110310

Zuboff, S. (2019). *The age of surveillance capitalism: The fight for a human future at the new frontier of power*. Public Affairs.

# Chapter 2
# Metaverse Stories

**Abstract** This chapter situates the metaverse in a long line of utopian and dystopian visions of a digital future in which being simultaneously 'here and elsewhere' is tied to claims about disruption and revolution. It does this by tracing the main narratives about virtual worlds and their historical origins—including their escapist visions of disembodiment and spatiality—back to twentieth- and early twenty-first century scientific and fiction literature, sci-fi movies, and corporate and political documents.

## 2.1 To Be Here and Elsewhere

The 'Metaverse' is only the most recent term to describe a long series of dreams, desires, and fantasies that have been shaping our world from the very beginning of human history. Media and communication scholars from the Toronto School (Innis, 1950; McLuhan, 1964) up to more recent contributions such as those by Bolter and Grusin (2000), Sherry Turkle (1995) and Boyd (2014), have repeatedly framed communication technologies as 'extensions' of our individual and social reality. History itself, as an academic discipline, began with writing, a technology that extends our thoughts and utterances on paper. As much as modernity is characterized by a "proliferation of infrastructures" (Peters, 2015, 31), it is also characterized by a proliferation of media and, concurrently, by technologies, devices, languages, and styles that allow us to project our lives, ideals, and imaginations in many different ways. In this context:

> VR and AR are among the latest additions to our complex media culture, which still includes everything from books and magazines to paintings, radio, television, and film — all competing for attention and status. In that competition, the producers and promoters of each form promise an experience that is unique and, in some way, better than other media — more compelling, more authentic, truer to life. (Bolter et al., 2021, 22)

Understanding the metaverse from a historical perspective means digging into the multiple processes—technological, political, and economic but also cultural—that have led up to its current conceptualization and development. Most historical accounts of the metaverse begin with the sci-fi novel *Snow Crash* written by Neil

© The Author(s) 2025

C. Hesselbein and P. Bory, *Infrastructures of Reality: Metaverse Stories, Spaces, Bodies*, PoliMI SpringerBriefs, https://doi.org/10.1007/978-3-031-97167-9_2

Stephenson, which inspired the rebranding of Facebook to Meta (Seebold & Nam, 2025). However, authors such as Bolter et al. (2021) trace the development of virtual and augmented reality back through the history of visual culture by focusing on the various technical and aesthetic features that were step by step developed and integrated into previous media technologies, thus exemplifying the process of remediation. Similarly, Evans (2019) stresses, in his book on the re-emergence of virtual reality, the relevance of nineteenth-century stereoscopic media while also underlining how technological constraints slowed down the development of VR, particularly during the 1980s. Other accounts have focused on how imaginaries of VR concurrently emerged with the rise of Silicon Valley and the countercultural movements of the 1960–80s (Lanier, 2017), or focused on so-called breakthrough devices such as the Oculus headset and the company's troubled path before its acquisition by Facebook in 2014 (Song, 2022).

In this chapter, rather than focusing on the recent development of virtual devices and environments, we will dig into a specific aspect of metaversal histories, namely narrativity. A fundamental presumption of this chapter is that how stories are told and presented to us matters profoundly. Whether backwards-looking or forwards-looking stories have a profound ability to shape how we interpret and give meaning to everything around us as well as ourselves. Indeed, the metaverse itself can be understood as a narrative construction. For instance, Zuckerberg's famous announcement at the Connect 2021 conference was much more about the foundation of a project rather than the presentation of a product. As Zuckerberg underlined during his announcement, the presentation of the metaverse was focused on "how we will build it together" (Meta, 2021) thus positioning the rise of the metaverse in the future rather than in the present.

However, the narrative of the future woven by Meta around the metaverse is not written from scratch; it is the result of a discursive construction that dialogues both explicitly and implicitly with the origins, development, and myths surrounding digital innovations and infrastructures over the past decades. In short, the metaverse imaginary draws on a much larger narrative about technology and/as progress. Metaverse narratives and stories are in themselves infrastructures of reality in that they act not only as forms of intermediation between different temporalities, but also as drivers of scientific and technological experimentation, economic investment, and as justifications for the anticipatory application of technological artefacts, especially those that still only exist on an imaginary level. As Wilhelm Schapp has argued, we are entangled (*verstrickt*) in stories (Schapp 1953); the stories we tell and share about past and future socio-technical developments shape our relationships with technology and the meanings we attribute to it, alongside and beyond our everyday interactions with and through technology. To understand the birth, or perhaps rather the conception of the metaverse, it is essential to trace back its theoretical and imaginary foundations in order to grasp the multiple perspectives that are shaping its current, though still 'magmatic' (Castoriadis, 1987) and uncertain definition.

In what follows below, we retrace and analyse several key narratives that illustrate the recurring fantasy of 'being here and elsewhere' through so-called 'new' media. These media have enabled multiple ways of being—or imaginings of being—here and elsewhere by holding out hopes for new forms of social, political, and economic interaction and organization in hybrid spaces, not to mention in virtual environments seemingly free from the limitations imposed by the materiality of bodies and things. Being here and elsewhere can mean either being in different places at once or moving rapidly back and forth from one place to another, but it can also mean being able to remain in the same space across different temporalities, inhabiting different bodies and identities. All media technologies tend to challenge our common perceptions of space and time. And each different combination of these technologies has specific characteristics that can stimulate the emergence of different narratives, often utopian or dystopian in character, about our future realities, our spaces, and our bodies with and through such media.

Below, we refer to several key narratives emerging from a plurality of sources, such as sci-fi novels and movies, political speeches and treatises, and corporate communications. In going through these sources, we examine linkages between the history of media and networking technologies with the expected and desirable—and sometimes undesirable—cultural and socio-technical outcomes as foreseen in the 'past futures' (Montross, 2015) or 'imaginary futures' (Barbrook, 2007) of sci-fi novelists, film directors, and other relevant actors from scientific, political and technological milieus.

For the sake of simplicity, this chapter employs a three-step periodization to structure the narratives and related imaginaries found in the history of the metaverse, starting from the first half of the twentieth century. However, the distinctions between these periods are somewhat artificial as elements of each can also be found in others. The first period focuses on the intersection or convergence of various media, such as motion picture, telephony and electricity; the second deals with the combination of digital computing and internet connectivity, while the last two sections focus on the current age and on the specific narrative of the future established by Meta in launching its metaverse project. Overall, this chapter presents a 'short history of metaverse stories', aiming to identify both continuities and changes in the dreams and imaginaries surrounding the metaverse while simultaneously underlining its defining features and sociocultural implications.

## 2.2 Dream Machines

One of the most famous and foundational myths in media historiography is the so-called 'train effect'. The legend is well known: the first screening of *L'Arrivée d'un train en gare de La Ciotat* by the Lumière brothers in 1896 is supposed to have triggered panic in the audience who believed that the train would crash through the frame, causing them to flee from the theatre. Although this event never really happened, as demonstrated by Bottomore (1999), the myth of the 'train effect' persists as one of the longstanding and most significant in media historiography. The reason is simple. The power of this myth is its symbolic efficiency, that is, it represents both the disruptive power of the cinematographic medium and its ability to radically change our perception of reality. In *Reality Media*, Bolter et al. (2021) begin their historical account of VR/AR with a second version of this myth. According to other studies on the origins of the train effect, rather than the first screening in 1896, it was the stereoscope experiments conducted by the Lumière brothers in a later screening of the same movie in the 1930s that triggered the awe and panic in the audience and thus gave rise to this myth. This second version emphasizes the important role of early technologies such as stereoscopic lenses, which could lend an illusion of 'greater depth' to images.

Although motion picture is widely acknowledged by media scholars as a radical change in visual culture, some nevertheless argue that stereoscopic cinema was preceded by other spectacular media, such as stereocards:

> Long before both the Lumiere Brothers and the Edison Manufacturing Company filmed their now iconic cinematic actualities of oncoming trains, for instance, many thousands of stereocards were produced that replicated a similar effect of a train emerging from an image towards the viewer (albeit without the movement that later characterized cinema and amazed the viewing public). Nor was this effect limited to trains, with stereographers capturing oncoming ships, zeppelins, artillery, horses, carriages and anything else that would accentuate the stereoscopic effect. (Gurevitch, 2013, 5)

Under the influence of the 'technological sublime' at the beginning of the twentieth century (Marx, 1964; Nye, 1996)—a collective sense of grandeur and wonder engendered by the achievements of the industrial revolution, which appeared capable of uniting nations under the umbrella of technological progress—science fiction writers started to combine elements of this 'new' media system, such as cinema, telegraphy, and photography, with the technological hallmarks of the industrial revolution, such as railway systems and electricity networks. Especially at the beginning of the century, the substitution of face-to-face interactions by other communication forms was a recurrent trope of sci-fi narratives. In the short novel *The Machine Stops* by Forster (1909), for example, future humans are described as living in underground dwellings in which it is possible to accomplish any routine activity, including communicating with thousands of other humans, through a dedicated telecommunication system provided by a god-like machine. In Forster's dystopian narrative, a mother is forced by her son to leave her home in order to take a train and prevent him from abandoning his dwelling. The room where she lives is depicted as a phantasmatic dream—or rather nightmare—of full automation:

> Then she generated the light, and the sight of her room, flooded with radiance and studded with electric buttons, revived her. There were buttons and switches everywhere— buttons to call for food, for music, for clothing. There was the hot-bath button, by pressure of which a basin of (imitation) marble rose out of the floor, filled to the brim with a warm deodorized liquid. There was the cold-bath button. There was the button that produced literature. And there were of course the buttons by which she communicated with her friends. The room, though it contained nothing, was in touch with all that she cared for in the world. (Forster, 1909, 3–4)

This depiction of the mother's room is the result of a pre-digital imaginary in which the mechanization of space can be understood as a precursor to digitalization. The myriads of buttons are a prototypical form of the click, and the dematerialized living spaces—the bare but hyper-functional room—are a precursor to the empty rooms in which future users live while navigating supposedly abundant virtual worlds with their digital devices. It is interesting to note how sensory elements play a specific role in this story. In Forster's narrative, natural light is frightening. The mother is shocked by the sunlight shining through the train window while travelling, and the main projection of reality in her ordinary life takes place through sounds and other sensory experiences rather than images. The extension or augmentation of reality thus occurs through sound media rather than visual media, which reflects the key role that technologies such as wireless telegraphy and radio were playing before the spread of cinema or television.

A few decades after the publication of Forster's novel, the rise of the motion picture would stimulate new visions and fantasies involving sensory and synesthetic experiences in which the reproduction of visual elements and their interconnection with other sensory phenomena played a crucial role. In 1940, Alfonso Bioy Casares published *The Invention of Morel*, a sci-fi novel that anticipated some of the constitutive elements of the current imaginary of the metaverse. In the novel, the protagonist seeks refuge from a mysterious illness but gets lost on an unknown island inhabited by a group of people who appear to be living in a peculiar, dreamlike state. The protagonist eventually learns that these people are not quite real but are rather the result of an experiment involving a machine that can recreate human beings and their lived experiences, capturing them in a kind of eternal loop. This loop is a recording of the everyday life of the inventor, Morel, and his friends and fellows; a machinic dream created to keep alive the memory, and especially the feeling, of his romantic relationship with a woman named Faustine. In a note addressed to future visitors of the island, Morel describes his machine as follows:

> With my machine a person or an animal or a thing is like the station that broadcasts the concert you hear on the radio. If you turn the dial for the olfactory waves, you will smell the jasmine perfume on Madeleine's throat, without seeing her. By turning the dial of the tactile waves, you will be able to stroke her soft, invisible hair and learn, like the blind, to know things by your hands. But if you turn all the dials at once, Madeleine will be reproduced completely, and she will appear exactly as she is; you must not forget that I am speaking of images extracted from mirrors, with the sounds, tactile sensations, flavours, odours, temperatures, all synchronized perfectly. An observer will not realize that they are images. And if our images were to appear now, you yourselves would not believe me. Instead, you would find it easier to think that I had engaged a group of actors, improbable doubles for each of you! (Bioy Casares, 1940, 88)

Morel's machine works thanks to a perpetual energy source powered by sunlight and sea tides; its magic lies first of all in reproducing reality due to the combination of the cinematographic medium with an unprecedented 'green' energy source. Morel's dream is a meta-reality that brings with it immortality, and in particular, the immortalization—which is both the perpetuation and the immobilization—of a social bond (in this case love) through a permanent projection of its lived experience. The island, the infinite engine, and the projection apparatus are a socio-technical assemblage representing a long-standing fantasy behind simulation technologies: overcoming our biological constraints (Davis, 2015). Another relevant classic of early twentieth-century sci-fi literature, Aldous Huxley's *Brave New World* (1932), similarly contains some important precursors to the synesthetic motion picture, but with more pessimistic and dystopic nuances compared to Bioy Casares. In Huxley's novel, one of the main forms of collective entertainment is a special type of movie theatre called a 'feelie' in which films can stimulate all the audience's senses. A feelie merges visual perception with tactile and olfactory sensations, enabling viewers to physically experience the emotions and sensations portrayed in the movie (feelie is a portmanteau of the terms feeling and movie). For Huxley, however feelies represent a form of entertainment in which real emotions, relationships, and even high art are replaced by superficial experiences, thus sacrificing individuality and authenticity in favour of new technocratic forms of control:

> That's the price we have to pay for stability. You've got to choose between happiness and what people used to call high art. We've sacrificed high art. We have the feelies and the scent organ instead. (Huxley, 1932, 151)

The dystopian narrative presented in Huxley's novel coincides with the critical theory of communication that was contemporaneously emerging among scholars from the Frankfurt School, especially through the work of Adorno and Horkheimer (1947). These critics of the culture industry, which they considered a driving force of so-called 'soft totalitarianism'—saw cinema, as well as all forms of broadcasting media, as the main weapon of mass distraction used to enculturate or indoctrinate people into capitalism and consumerism. The role of media in establishing and maintaining cultural hegemony are longstanding topics that have been recently taken up by critical scholars such as Žižek (2000) and Fuchs (2020), among others. Even more recent theories, such as Cory Doctorow's concept of "enshittification" (2025)—i.e. the planned, gradual deterioration of online platforms, services, and products—go in a similar direction, underscoring the false sense of freedom associated with consumers' choices and the concentration of power in the hands of 'Big Tech' companies and their marketing and profiling strategies.

Science fiction literature from the first half of the twentieth century thus offers an interesting array of visionary narratives about the future of media technologies, laying the foundation for several recurring topics as well as critical perspectives on the relationship between media, reality, and society. However, in all these past futures, a genuine convergence between communication technologies and multisensory experiences, essential for the current conceptualization of the metaverse, was not yet being contemplated. In a similar vein, other cultural works, such as Fritz Lang's *Metropolis* (1927), acted as a bridge between the ancient myth of Prometheus, Mary Shelley's *Frankenstein*, and the rise of artificial intelligence, thus anticipating some key themes in the later emergence of AI. One pertinent example of this is the hubris of scientists seeking to create a being as intelligent as humans, if not superior to them, thus stealing from God the act of creation just as Prometheus had stolen fire from the Greek gods. However, it was only in the 1960s, concurrently with the rise of digital computing and computer networks, that ideas about connectivity and the dematerialization of experience began to be combined in Western sci-fi narratives as well as the broader sociotechnical imagination.

## 2.3 Digital Infrastructures as Stories

The biggest obstacle to the realization of the metaverse, particularly as conceived by Meta today, is probably the so-called interoperability issue, that is, the creation of a network of interconnected worlds through which one can navigate freely without the limits imposed by different and incompatible standards, devices, and operating systems. In the history of digital media and infrastructures, similar concerns about issues of interoperability between different computer systems emerged immediately after World War II, but these were fully addressed only decades later with the spread of the internet and especially with the birth of the World Wide Web in the mid-1990s.

Between the late 1940s and the 1970s, the great popularity of science fiction—think of authors such as Isaac Asimov, Philip K. Dick, and Kurt Vonnegut—went hand in hand with the proliferation of futurist narratives spread by computer scientists and cyberneticians, who, during the first phases of the digital revolution (Balbi, 2024) started to conceptualize what relations and interactions between humans, computers, and information infrastructures might look like (Bory, 2020). From Vannevar Bush's Memex (1945), Joseph Licklider's intergalactic network (1960), Douglas Engelbart's augmented intelligence (1962), and the hypertextual dream of Xanadu by Ted Nelson (1967) to the subsequent development of hypertext, Arpanet, and then of the internet, all these developments took place alongside the formulation of a wide variety of futuristic scenarios. More precisely, in these decades, the challenge of achieving interoperability between communication systems fed into at least two diverging narratives concerning freedom in the digital future. On the one hand, computer networks were conceptualized as technologies that would lead to new forms of collective intelligence and knowledge sharing, and in the most extreme version of this narrative, to a complete separation between the 'real' and a new and better digital world. On the other hand, in another narrative, created by Western governments and later reproduced during the Al Gore vice-presidency in the 1990s, the internet and the Web were portrayed as new highways that would connect a global market in a world free from war and geopolitical conflicts. This second narrative also influenced several local political movements, such as civic networks, as we discuss further below.

The first narrative is symbolized by the concept of cyberspace, first introduced in the cyberpunk novel *Neuromancer* by Gibson in 1984. In the 1990s, cyberspace was further conceptualized as an entirely new interconnected system that could serve as the infrastructure for a new information universe. This reformulation, however, opened up a series of utopian and dystopian scenarios. For example, the perception that humanity might be able to transcend the material constraints of geography, politics, and corporeality led to suggestions for the creation of a so-called collective intelligence that would surpass that of the single units of which it consisted (Lévy, 1997) as well as to anarchic ideas about a self-regulated virtual world in which everyone would be free and able to express themselves as they might wish. However, such dreams of collective intelligence or a society freed from political and economic constraints would soon be crushed by the rise of surveillance capitalism, which would make data and interoperable infrastructures its main source of value.

The culmination of the narratives surrounding cyberspace in the 1990s—a space freed from space—is the famous *A Declaration of the Independence of Cyberspace* shared on the Web by John Perry Barlow during the World Economic Forum at Davos in 1996:

> Governments of the Industrial World, you weary giants of flesh and steel, I come from Cyberspace, the new home of Mind. On behalf of the future, I ask you of the past to leave us alone. You are not welcome among us. You have no sovereignty where we gather. […] Cyberspace consists of transactions, relationships, and thought itself, arrayed like a standing wave in the web of our communications. Ours is a world that is both everywhere and nowhere, but it is not where bodies live. We are creating a world that all may enter without privilege or prejudice accorded by race, economic power, military force, or station of birth. We are creating a world where anyone, anywhere may express his or her beliefs, no matter how singular, without fear of being coerced into silence or conformity. Your legal concepts of property, expression, identity, movement, and context do not apply to us. They are all based on matter, and there is no matter here. (Barlow, 1996)

The rhetorical force of Barlow's *Declaration* lies in its combination of specific adjectives and metaphors. On the one hand, the giants made of flesh and steel, on the other, an immaterial mind independent of physical constraints. In cyberspace, according to Barlow, where there is no matter and where reality is "a standing wave in the web of communications", there is no material power either.

Barlow's declaration was one of the first texts to go viral on the Web. Importantly, the rhetorical and material dimensions of this declaration were deeply intertwined. Until the late 1990s, networking infrastructures could handle only small amounts of data. The dematerialization of geographical distance could mainly be achieved through a so-called light and poor medium, such as text, rather than a more data-heavy medium such as image or sound. These limitations in terms of bandwidth, as noted by virtual-reality pioneer Jaron Lanier (2017), contributed to the slowdown of heavier technologies, such as VR, which would consequently remain in the realm of sci-fi narratives and video arcades.

As a consequence, text was fundamental to the development of early cyberspace and virtual identities. In particular, beside developing their avatars, early users created and fostered new online identities in chatrooms and early MMPORGs (Massively multiplayer online role-playing games), literally writing themselves down via instant messages or asynchronous conversations with others and thus weaving their relationships with these so-called second selves (Turkle, 2005). According to Turkle, in the digital age technology became a new "architect of our intimacies" (Turkle, 2005). While the first avatars allowed for early explorations in reshaping identity, it was online relationships that were seen as a means for exploring the self in a more authentic manner than compared to 'real' physical interactions. During this period, the online creation of avatars and fictional characters was seen as a way to project desires about one's body, gender, and social identity (Casilli, 2010). In cyberspace one could embody not only someone else – if not something else, such as a monster or a fantastical creature in roleplaying games – but also explore more authentic selves in ways far removed from the restrictions imposed by social conventions or expectations and stigmas. These new digital spaces had a strong playful connotation and, thanks to anonymity and the lack of institutional and jurisdictional oversight, were perceived as neutral environments that were less likely to involve social judgment or have reputational repercussions on the 'real' social life of users. However, enthusiasm for these new mediated socialities and experimental identities was accompanied by fear about individual and collective loss, particularly in relation to the so-called myth of immersivity according to which virtual spaces risked turning into cages from which individuals could no longer escape, or would no longer want to escape. This fear of being enclosed has ancient origins, starting from Plato's cave to the total cinema (Bazin, 2004) and reappears in the harsh critiques of simulation and virtuality formulated by French theorists such as Baudrillard (1981) and Virilio (2000), among others.

Curiosity and uncertainty about the nature of virtual practices and the future of the digital society led to the production of a wealth of sci-fi movies between the 1980s and 1990s—a golden age for the genre of sci-fi cinema—dedicated to exploring the relationship between the body, memory, identity, technology, and the emergence of augmented and virtual reality devices and environments. From the video game immersion of Steven Lisberger's *Tron* (1982) to the cyberpunk visions of the console-body in David Cronenberg's *Existenz* (1999) and the relivable memories of the SQUID (Superconducting QUantum Interference Device) in Kathryn Bigelow's *Strange Days* (1995), up to the scientistic hubris of Brett Leonard's *The Lawnmower Man* (1992) and the grotesque virtual game character of Gabriele Salvatores' *Nirvana* (1997), cinematic science fiction over these two decades visually represented and interrogated the various fears and pleasures that digital immersion, including diving into our own brains and memories, might produce. Such movies and the imaginaries of virtual reality they contained have already received much scholarly attention (e.g. Chan, 2014). Some lesser-known cinematic productions, however, have explored other interesting aspects of virtual immersion, particularly from a sociological point of view. For example, *Thomas in Love* by Pierre Paul Renders (2000) in which the protagonist, a thirty-year-old man named Thomas, has never left his home since childhood. Like the mother in Forster's novel, Thomas's life is in a state of permanent connection in which the screen has replaced Forster's buttons; he goes to the doctor, orders food, and does everything he needs to live online while sitting in his dark room. His agoraphobia makes Thomas incapable of dealing with the outside world or having any human contact, his life is on the internet. Films such as *Thomas in Love* were released at the turn of the millennium when the diffusion curve of the internet and personal computing was rising globally. It is at this juncture that the Wachowski siblings released their movie *The Matrix* (1999), which combined dystopian aspects of the immersion narrative with ideas about the advent of so-called strong AI, and subsequently resulted in one of the most successful narratives in the history of sci-fi cinema. *The Matrix* is so far the only imaginary world in which the concept of framelessness is realized and represented (Carter & Egliston, 2024), as the main medium between the human characters and the virtual world they live in is located in their brain; there is no means for the characters to distinguish between what is real or simulated.

One year before the release of *The Matrix*, another movie dealing with an alternative form of reality attracted the attention of the general public as well as that of critical scholarship. Relying on the 'old' medium of television, *The Truman Show* (1998) by Weir contrasted the dystopia of virtual reality as represented by *The Matrix* with a more plausibly realistic but equally fictional setting in which reality is mediated by human actors rather than by AI-controlled simulations. The protagonist Truman ('the only true man') unwittingly lives in a Big Brother-type show where all the other people are actors playing their part and co-running the television show that has been broadcasting his entire life from birth. *The Truman Show* forcefully challenges the idea that virtual reality and fictional lives must be mediated by technology. We will return to this point later, after exploring the emergence of another narrative in the 1990s—arising concomitantly with the cyberspace narrative—which connects the realm of the virtual with civic participation rather than individual immersion.

## 2.4  Netizens and Early Digital Twins

The conceptualization of the internet as a free and anonymous space for experimenting with the self and for recreating new forms of virtual sociality was accompanied, especially during the late 1980s and early 1990s, by the emergence of another narrative in which the virtual realm was imagined as an enhancement of the public sphere rather than as its replacement with an unregulated cyberspace. While Western countries were investing huge material and symbolic resources in promoting the so-called information superhighways, the internet and especially the Web seemed to provide the foundation for a new democratic arena in which citizens could directly partake in political decisions. This chapter is not the place to dig into the history of the transformation of these so-called virtual communities (Rheingold, 1993) and their role in the rise of the so-called network society (Castells, 2000). However, it is important to highlight the narratives of civic participation that emerged in parallel to the alleged dematerialization of digital space. This is particularly the case because the same digital tools that were once imagined as boosters of grassroots activism and collective participation in politics—at first internet connectivity but later also smartphones and social media—are today often criticized as the causes of the people's detachment from the *res publica* rather than enablers of collective politics.

In the 1990s, imaginaries of a new active netizenship (Hauben and Hauben, 1997) were emerging, which were posited, with variable degrees of success, as important alternatives to the framing of the cyberspace as a virtual world separated from concrete reality. It is in this context that the first civic networks were born as platforms of intermediation and co-participation between national governments, local municipalities, and citizens, especially in several major European cities. One of the first civic networks was De Digitale Stad (DDS), launched in 1994 in Amsterdam (Alberts et al., 2017), but other important civic initiatives, such as Bologna's network Iperbole, created in 1993, were also set up to provide free internet access to citizens for the first time in Europe (Bory, 2019). Such experiments in fostering active citizenship through the internet involved different stakeholders, such as activists, local governments, and prominent philosophers and computer scientists, as well as common citizens who contributed to the construction of a new online public sphere (the 'netsphere'). These civic initiatives can be compared to the emergence of previous grassroots initiatives, such as pirate radio stations and early computer hobbyist communities. Notably, these European civic networks combined multiple visual and textual elements that anticipated the birth of the so-called digital twins (see next chapter), such as navigable maps of cities as well as virtual squares where users could access a wide range of services, read news items, and find practical information about their municipalities' activities and policies (Tambini, 1998). These services were made available alongside online discussion groups in which citizens could collectively debate policies and thus participate in decision-making processes, exchange goods and professional services, and provide feedback on the efficacy of actions taken by governance institutions at local and regional levels.

Amsterdam's DDS has been preserved in the UNESCO digital heritage platform and is described as follows:

> On January 15, 1994, De Digitale Stad (DDS, The Digital City) opened its virtual gates. It empowered users to be "citizens" or "netizens" of a digital environment and enter the then largely unknown world of the Internet. DDS imagined a community-driven, commons-based internet, far remote from today's internet whose operation is dictated by a small group of big tech companies. (UNESCO, 2023)

DDS employed the architectural metaphor of 'town squares' to guide citizens in finding information on the city's website and included gamification features that allowed users to configure their own, quite basic, avatars that could be used to navigate the website and participate in online activities. Here, in contrast to the cyberspace narrative in which the online 'world' was seen as detached from real world politics, the internet was conceptualized by civic network movements as a new space for enhancing public participation and for recreating the publicness and openness of 'real life' spaces. Practices of online "square-hopping" (Francissen & Brants, 1998) fostered a feeling of co-presence and co-participation among citizens in this new virtual square, and, moreover, also encouraged face-to-face meetings and encounters in physical spaces. Unlike social media platforms—defined by the critical scholar Geert Lovink, one of the founders of DDS, as "networks without a cause" (Lovink, 2011)—civic networks were created and narrated as tools for re-building and re-enacting direct democracy. These civic networks also served as precursors to a series of later and larger-scale political movements (such as the 5 Star Movement in Italy and the Pirate Parties in Northern and Eastern Europe) that would make the network-citizenship bond a core point of their political and organizational identity. While Neil Stephenson wrote about a future metaverse monopolized by 'Big Tech' companies and dominated by economic elites, in the 1990s the internet was still seen as a tool to speed up direct democracy as well as an infrastructure to facilitate other activities such as teleworking and telemedicine.

However, network technology of the 1990s was not sufficiently advanced for the realization of such dreams of democratic togetherness or of a truly liveable virtuality. These virtual worlds were still relatively poor in terms of internet connectivity, graphics, and technological sophistication. Nevertheless, they were also rich in the sense that they embodied and inspired collective imaginations and expectations of the future. Soon, however, these political ideals and expectations would be betrayed, only to be taken up again a few decades later, yet this time shaped by a different vision of a sociotechnical future espoused by one particular leading company.

## 2.5   Birth and Death of the Metaverse

The speed with which computing power and internet connection have developed from the late 1990s to today has facilitated the spread of increasingly sophisticated and complex technologies, software, and virtual environments, but these, however, require a long process of domestication on the part of users that is not only technical but also symbolical, physical, and psychological. Although, as Carter and Egliston have aptly shown (2024, 5–28), companies such as Sega and Nintendo failed several times in their attempts to make VR a global phenomenon, this is to a great extent due to psycho-physical effects such as motion sickness as well as users' hesitancy to be fully immersed in multisensorial virtual landscapes and experiences. After all, calibrating bodily senses and technological standards at the collective level is a delicate and slow process. For example, it took decades to make 3D visuals familiar and user-friendly, even for enthusiastic expert users such as gamers, via first-person games played on conventional screens. Only recently has this process of familiarization and habituation started for mixed-reality headsets. From a technological perspective the history of the metaverse is part of the long history of visual media, ranging from pictorial art to 3D virtual environments, and part of a process in which technologies of various kinds transformed from the realm of the sublime to the banal (Mosco, 2005) over the span of decades, if not centuries.

The history of the metaverse also involves the decline and (false) resurrection of a narrative of togetherness (Simonson, 1996) that posits media technologies as revolutionary and democratic infrastructures. Notably, the beginning of the twenty-first century marked the end of both the utopian narrative of cyberspace and the democratic narrative of digital citizenship. Both myths or sublime visions (Mosco, 2005), once subjected to scalability by 'Big Tech' companies, have succumbed to the dynamics of digital capitalism. The "net delusion" (Morozov, 2011) of the Arab Spring movements, as well as the rise and fall of the "people of the net" (Natale & Ballatore, 2014)—both once hailed as important emerging political movements—are striking examples of how internet technology has failed to replace the infrastructures of political intermediation. But above all, these developments demonstrate how digital infrastructures and media—as well as the narratives surrounding them—have not been able to replace the role that grand narratives and political ideologies have played in holding societies together. For decades, several cyber-communitarian movements—most of them from the left of the political spectrum—regarded technologies such as the internet and computing as vehicles for advancing egalitarian and democratic processes. More recently, however, new systems and technologies, such as the metaverse and AI, have come to represent civilizational progress in themselves.

Interestingly, one of the main architects behind the demolition of the myth of cyberspace was Mark Zuckerberg. In 2004, Linden Lab's Second Life seemed to open the doors to the creation of a truly interconnected virtual world. The growth of computational power and internet connectivity gave impetus to a new virtual realm whose slogan was, "Your world, your imagination", which triggered the interest of social scientists and ethnographers trying to understand this new environment and form of sociality consisting of multicoloured landscapes and outlandish avatars in which U2 played its first virtual concert and where users were creating and shaping their second lives. We will not elaborate here on the main features of this well-known virtual world, which, despite criticism, re-emerges in some narratives as the forerunner to if not the only true metaverse (Au, 2023). What is interesting to note from a historical and narratological point of view is that the decrease in interest in Second Life during the late 2000s coincided with the global diffusion of social media platforms, especially with the birth of Facebook and particularly its rise from 2007 onwards. What Facebook's slogan of making the "world more open and connected" (Hoffman et al., 2018) made clear—together with similar slogans such as YouTube's "broadcast yourself"—was that the greatest source of inspiration (and financial profit) in the digital age was not fantasy or imagination but reality. Rather than imaginary worlds, creative avatars, or new social relations, the emergence of social media and 'reality' television placed self-broadcasting and ordinary lives at the centre of the internet as well as the broader media landscape.

There are many factors lying behind the paradigm shift that affected the entire digital landscape, and one of them is certainly financial. Frankly, big tech companies realized that real selves were much more profitable than imaginary ones. Our faces, clothes, voices, tastes, and choices have become the main resource of the digital economy in the form of data. Starting from the birth of the Google search engine in the late-1990s, through to the social media platforms and the emergence of Web 2.0 in the 2000s and the current rise of generative AI, personal data has been front and centre. The data 'Big Tech' companies collected over time on users and their networks, also thanks to the broadcasted selves mentioned just above, have become a financial gold mine and a fundamental resource for consolidating and perpetuating their dominant position in the global market. However, these companies nevertheless still have to cultivate and maintain their symbolic power in addition to their economic and technological powers. 'Big Tech' companies must appear as cutting-edge and ahead of the times in order to maintain the interest of investors. Corporate rebranding strategies point exactly in this direction; companies such as Facebook and Google change their names, make promises of progress and radical technological change, and pretend to open the market doors to new players and competitors. But their main goal is maintaining the current power (un)balance and their dominant position and control over the digital market and the data economy. Historically, powerful actors have constantly needed to demonstrate their ability to adapt to the 'new', if not just to anticipate and shape the conditions for progress and change; contemporary corporate giants are no exception to this.

## 2.6   2021—The Resurrection

Digital media history is rich in mythopoetic sources such as governmental documents, academic papers, manifestos, and letters from CEOs to investors, among other examples. Some of these documents have been chronicled as key turning points in history. Think of Alan Turing's seminal paper (1950) that proposed the famous Turing Test and how this started a never-ending debate on the meaning of 'artificial intelligence', or the aforementioned *A Declaration of the Independence of Cyberspace* by Barlow (1996) and its juxtaposition of national states with anarchic networks, or how the *Computerization of Society* by Nora and Minc (1978) fed into European and US discourse on the need for digitization and high-speed infrastructures in the 1980s and 1990s, or how the famous *An Open Letter to Hobbyists* written by Gates (1976) heralded in the age of proprietary software and eventually led to the creation of the neoliberal software marketplace.

In relation to such canonical texts, and well before the launch of Meta in 2021, Mark Zuckerberg had already employed the medium of the letter to glorify and narrate the past and the future of his company as well as its ostensibly key role in society at large. In 2012, the CEO of Facebook wrote a famous letter to investors that sought to gain their trust by underscoring the company's outstanding accomplishments as well as its contemporary and future achievements:

> At Facebook, we're inspired by technologies that have revolutionized how people spread and consume information. We often talk about inventions like the printing press and the television - by simply making communication more efficient, they led to a complete transformation of many important parts of society. They gave more people a voice. They encouraged progress. They changed the way society was organized. They brought us closer together. [...] Today, our society has reached another tipping point. We live at a moment when the majority of people in the world have access to the internet or mobile phones - the raw tools necessary to start sharing what they're thinking, feeling and doing with whomever they want. Facebook aspires to build the services that give people the power to share and help them once again transform many of our core institutions and industries. (Zuckerberg, 2012)

Inspired by other prominent 'tech' CEOs such as Steve Jobs, Zuckerberg placed his company in a direct line with the historical development of other media, particularly when claiming that Facebook was the networking infrastructure that had wired, and would rewire, social, economic and cultural connections worldwide, just as previous revolutionary technologies such as the printing press and television had "led to a complete transformation of many important parts of society". This narrative fits perfectly with the concept of corporational determinism (Natale et al., 2019), that is, the recurring strategy of digital media corporations to narrate themselves as the main agents of societal and technological change. The appropriation of the imaginary of other characters and communities in 'tech' and media history is also illustrated by the subtitle of Zuckerberg's letter, "The hacker way", which suggests that the company embodied the values of hacker culture. However, the hacker movement has long been critical of closed systems such as Facebook's platforms as well as companies employing end-to-end processes such as Apple.

Almost ten years later, in his letter of October 2021, Zuckerberg appeared to reenact a similar narrative:

> We are at the beginning of the next chapter for the internet, and it's the next chapter for our company too. [...]The metaverse will not be created by one company. It will be built by creators and developers making new experiences and digital items that are interoperable and unlock a massively larger creative economy than the one constrained by today's platforms and their policies. [...] To reflect who we are and the future we hope to build, I'm proud to share that our company is now Meta. Our mission remains the same — it's still about bringing people together. Our apps and their brands aren't changing either. We're still the company that designs technology around people. But all of our products, including our apps, now share a new vision: to help bring the metaverse to life. And now we have a name that reflects the breadth of what we do. [...] Ours is a story that started in a dorm room and grew beyond anything we imagined. [...] We have built things that have brought people together in new ways. We've learned from struggling with difficult social issues and living under closed platforms. Now it is time to take everything we've learned and help build the next chapter. [...] I'm dedicating our energy to this — more than any other company in the world. If this is the future you want to see, I hope you'll join us. The future is going to be beyond anything we can imagine. (Meta, 2021)

Read through an historian's lens, Zuckerberg's letter looks like a hodgepodge of the narratives and the past futures presented and analysed in this chapter. Perhaps the most contradictory aspect of his letter is that the end of the platform economy and the emergence of a new interoperable and horizontal system—the metaverse—will supposedly be driven by one of its leading players, if not the archetype of the platformization of the internet (Bucher, 2021; Plantin et al., 2018). Moreover, situating the humble origins of the great company in a college dorm room reiterates an old cliché of Silicon Valley, namely that of the genius who emerges from below, pushing his revolutionary discovery in the service of the entire society, but ignores the fact that a dorm room at Harvard is not that low on the social scale of the wealthiest country in the world. Furthermore, Zuckerberg's Metaverse is very similar to the initiative that he contributed to undermining by founding Facebook in the mid-2000s—i.e. Second Life and virtual worlds—but his version is emptied of the ideals that such metaverses once embodied, such as the search for new forms of community and self-exploration that surrounded virtual-reality technologies and digital worlds of the 1990s and early 2000s. From the point of view of collective imaginaries, the resurrection of the metaverse by Meta is characterised by a degree of degradation that is akin to the zombies found in fantastical literature; creatures resurrected by the force of technology but that have lost their humanness in the process. Rather than the sci-fi imaginary presented by Stephenson in *Snow Crash*, which challenges the idea of a horizontal and 'democratic' metaverse—or the rhetoric of technological and perceptual revolution surrounding VR technologies—Meta's narrative of the Metaverse is buttressed only by its economic power. The ten thousand jobs in Europe promised by Meta for the development of metaverse technologies, its influence over the news media as one of the largest global platforms, the acquisition of Oculus and the dominance of the Quest headset in VR hardware, and its ownership of the most widely used social media platforms and messaging apps in the Western world; these are the reasons that have made the metaverse and its entanglement with Meta one the main topics of public, political, economic and academic debate, this book included. Even though major Hollywood productions such as *Avatar* (Cameron, 2009) and *Ready Player One* (Spielberg, 2018) reproduced dominant imaginaries surrounding immersive technologies, such narratives were more inspired either by questions around the existence of other species, such as in the case of *Avatar*, or by interest in exploring the role of public figures and cultural references from the last decades of the twentieth century. In *Ready Player One*, for example, the character of the proto-Steve Jobs and nostalgia for role-playing games of the 1980s–1990s functioned more as representations of the "future that used to be" (Barbrook, 2007) rather than of the future imagined today by companies such as Meta.

We should not deny that it is possible that virtual technologies and environments are rapidly evolving and could well become part of our daily lives in the short or long term, in different ways, and potentially with more or less desirable outcomes. Just as it is necessary to examine narratives of the metaverse imposed from above by 'Big Tech' companies, it is equally necessary to highlight other visions of the metaverse's potential future as well as to uncover missing perspectives that mainstream narratives can conceal.

An emergent body of scholarship argues that hype is replacing hope in contemporary imaginaries (Galanos & Stewart, 2024). If we consider hope exclusively as a collective phenomenon aimed at the common good, it is probably true that technological hype is now supplanting the dreams of togetherness that were associated with media of the past such as radio, television, and the internet. But hope provides nourishment to ideals of progress found in democratic societies as well as in companies. The 'launch' of the metaverse, however, represents specific corporate hopes and imaginaries in which embodied experience and virtual environments are tools to sustain the delusions of grandeur of a small group of actors rather than tools to support broader democratic ideals or provide to social needs.

During an interview right after the launch of Meta in 2021, Zuckerberg claimed that the metaverse is "an embodied internet and you're in it and *the atomic unit is you* have your avatar and your digital goods. [...] I think that's an architecture that should be fundamentally more amenable to interoperability" (Heath, 2021, emphasis added).

This short excerpt highlights how single users are conceptualized as "the atomic unit" of a metaverse in which individuals, and particularly their bodies, are part of the "architecture" of the virtual world envisioned by Meta. More than one century ago, however, understanding and mitigating the atomization of society was considered one of the foremost challenges of modernity, an issue that led to the birth to the social sciences and that stimulated seminal reflections by authors such as Ferdinand Tönnies and Georg Simmel on the fragmentation of society and the decline of social reciprocity. Zuckerberg's imaginary of a metaverse—consisting of billions of connected monadic elements linked "alone together" (Turkle, 2011), using their datafied bodies to work and play in quantified spaces—brings this genealogical question back to the heart of the social sciences and humanities. It is precisely for this reason that hidden or alternative narratives and even counter-narratives should be identified and highlighted; to understand what the metaverse is or can be, but also what it will do to us, our bodies, and our social and living spaces.

# References

Adorno, T. W., & Horkheimer, M. (1947). *Dialektik der Aufklärung*. Querido Verlag.
Alberts, G., Went, M., & Jansma, R. (2017). Archaeology of the Amsterdam Digital City: Why digital data are dynamic and should be treated accordingly. *Internet Histories, 1*(1–2), 146–159. https://doi.org/10.1080/24701475.2017.1309852
Au, W. J. (2023). *Making a metaverse that matters: From snow crash & second life to a virtual world worth fighting for*. John Wiley & Sons.
Balbi, G. (2024). *The digital revolution: A short history of an ideology*. Oxford University Press.
Barbrook, R. (2007). *Imaginary futures: From thinking machines to the global village*. Pluto Press.
Barlow, J. P. (1996). *A declaration of the independence of cyberspace*. Retrieved March 15, 2025, from https://www.eff.org/it/cyberspace-independence
Baudrillard, J. (1981). *Simulacres et Simulation*. Éditions Galilée.
Bazin, A. (2004). *What is cinema?* Vol. I. University of California Press.
Bigelow, K. (1995). *Strange days*. Lightstorm Entertainment.
Bioy Casares, A. (1940). *The invention of morel*. The New York Review Books.
Bolter, J. D., Engberg, M., & MacIntyre, B. (2021). *Reality media: Augmented and virtual reality*. MIT Press.
Bolter, J. D., & Grusin, R. (2000). *Remediation: Understanding new media*. MIT Press.
Bory, P. (2019). The Italian network hopes: Rise and fall of the Socrate and Iperbole project in the Mid-1990s. *Internet Histories, 3*(2), 105–122. https://doi.org/10.1080/24701475.2019.1596407
Bory, P. (2020). *The internet myth: From the internet imaginary to network ideologies*. University of Westminster Press. https://doi.org/10.16997/book48
Bottomore, S. (1999). The panicking audience? Early cinema and the 'train effect.' *Historical Journal of Film, Radio and Television, 19*(2), 177–216. https://doi.org/10.1080/014396899 100271
Boyd, D. (2014). *It's complicated: The social lives of networked teens*. Yale University Press.
Bucher, T. (2021). *Facebook*. John Wiley & Sons.
Bush, V. (1945). As we may think. *The Atlantic Monthly, 176*(1), 101–108.
Cameron, J. (2009) *Avatar*. 20th Century Fox.
Carter, M., & Egliston, B. (2024). *Fantasies of virtual reality: Untangling fiction, fact, and threat*. MIT Press.
Casilli, A. A. (2010). *Les Liaisons Numériques. Vers une Nouvelle Sociabilité?* Seuil.
Castells, M. (2000). *The rise of the network society*. John Wiley & Sons.
Castoriadis, C. (1987). *The imaginary institution of society*. MIT Press.
Chan, M. (2014). *Virtual reality: Representations in contemporary media*. Bloomsbury Publishing USA.
Cronenberg, D. (1999). *Existenz*. The Movie Network.
Davis, E. (2015). *TechGnosis: Myth, magic, and mysticism in the age of information*. North Atlantic Books.
Doctorow, C. (2025). *Enshittification: Why everything suddenly got worse and what to do about it*. MCD.
Engelbart, D. C. (1962). Augmenting human intellect: A conceptual framework. In R. Packer, & K. Jordan (Eds.), *Multimedia. From wagner to virtual reality* (pp. 64–90). WW Norton and Company.
Evans, L. (2019). *The re-emergence of virtual reality*. Routledge.
Forster, E. M. (1909). *The machine stops*. The Oxford and Cambridge Review.

Francissen, L., & Brants, K. (1998). Virtually going places. In R. Tsagarousianou, D. Tambini, & C. Bryan (Eds.), *Cyberdemocracy: Technology, cities and civic networks* (pp. 18–40). Routledge.

Fuchs, C. (2020). *Communication and capitalism: A critical theory.* University of Westminster Press. https://doi.org/10.16997/book45

Galanos, V., & Stewart, J. (2024). Navigating AI beyond hypes, horrors and hopes: Historical and contemporary perspectives. In A. Ponce del Castillo (Ed.), *Artificial intelligence, labour and society* (pp. 27–46). ETUI.

Gates, B. (1976, January 31). An open letter to hobbyists. *Homebrew Computer Club Newsletter, 2*(1). Retrieved March 15, 2025, from https://www.digibarn.com/collections/newsletters/hom ebrew/V2_01/index.html

Gibson, W. (1984). *Neuromancer.* Ace Books.

Gurevitch, L. (2013). The stereoscopic attraction: Three-dimensional imaging and the spectacular paradigm 1850–2013. *Convergence, 19*(4), 396–405. https://doi.org/10.1177/135485651 3494175

Hauben, M., & Hauben, R. (1997). *Netizens: On the history and impact of Usenet and the internet.* IEEE Computer Society Press.

Heath, A. (2021, 28 October). Mark Zuckerberg on why Facebook is rebranding to Meta. *The Verge.* https://www.theverge.com/22749919/mark-zuckerberg-facebook-meta-company-rebrand

Hoffmann, A. L., Proferes, N., & Zimmer, M. (2018). "Making the world more open and connected": Mark Zuckerberg and the discursive construction of Facebook and its users. *New Media & Society, 20*(1), 199–218. https://doi.org/10.1177/1461444816660784

Huxley, A. (1932). *Brave new world.* Vintage Books.

Innis, H. (1950). *Empire and communications.* Oxford University Press.

Lang, F. (1927). *Metropolis.* Babelsberg Studio for Universum Film A.G. (UFA).

Lanier, J. (2017). *Dawn of the new everything: A journey through virtual reality.* Bodley Head.

Leonard, B. (1992). *The lawnmower man.* Allied Vision, Fuji Eight Company Ltd., Lane Pringle Productions.

Lévy, P. (1997). *Collective intelligence: Mankind's emerging world in cyberspace.* Perseus Books.

Licklider, J. C. R. (1960). Man-computer symbiosis. *IRE Transactions on Human Factors in Electronics, HFE-1*(1), 4–11.

Lisberger, S. (1982). *Tron.* Walt Disney Pictures.

Lovink, G. (2011). *Networks without a cause: A critique of social media.* Polity Press.

Marx, L. (1964). *The machine in the garden: Technology and the pastoral ideal in America.* Oxford University Press.

McLuhan, M. (1964). *Understanding media: The extensions of man.* McGraw-Hill.

Meta (2021). *Founder's letter, 2021.* Retrieved March 15, 2025, from https://about.fb.com/news/2021/10/founders-letter/

Montross, S. J. (Ed.) (2015). *Past futures.* MIT Press.

Morozov, E. (2011). *The net delusion.* Perseus Books.

Mosco, V. (2005). *The digital sublime: Myth, power, and cyberspace.* MIT Press.

Natale, S., & Ballatore, A. (2014). The web will kill them all: New media, digital utopia, and political struggle in the Italian 5-Star Movement. *Media, Culture & Society, 36*(1), 105–121. https://doi.org/10.1177/0163443713511902

Natale, S., Bory, P., & Balbi, G. (2019). The rise of corporational determinism: Digital media corporations and narratives of media change. *Critical Studies in Media Communication, 36*(4), 323–338. https://doi.org/10.1080/15295036.2019.1632469

Nelson, T. H. (1967). Getting it out of our system. In G. Schechter (Ed.), *Information retrieval: A critical review* (pp. 191–210). Thompson Books.

Nora, S., & Minc, A. (1978). *L'Informatisation de la Société.* Documentation Française.

Nye, D. E. (1996). *American technological sublime.* MIT Press.

Peters, J. D. (2015). *The marvelous clouds: Toward a philosophy of elemental media.* University of Chicago Press.

Plantin, J. C., Lagoze, C., Edwards, P. N., & Sandvig, C. (2018). Infrastructure studies meet platform studies in the age of Google and Facebook. *New Media & Society, 20*(1), 293–310.

Renders, P.P. (2000). *Thomas in love.* Entre Chien et Loup, JBA Production, Radio Télévision Belge Francophone (RTBF).

Rheingold, H. (1993). *The virtual community: Finding connection in a computerized world.* Addison-Wesley Longman Publishing Co.

Salvatores, G. (1997). *Nirvana.* Vittorio Cecchi Gori.

Schapp, W. (1953). *In Geschichten Verstrickt.* Verlag Richard Meiner.

Seebold, P., & Nam, C. S. (2025). A brief history of the Metaverse: From early beginnings to the cutting edge. In S. Chang, D. Nam, & H. J Song (Eds.), *Human-centered metaverse* (pp. 1–9). Morgan Kaufmann.

Simonson, P. (1996). Dreams of democratic togetherness: Communication hope from Cooley to Katz. *Critical Studies in Mass Communication, 13*(4), 324–342. https://doi.org/10.1080/152 95039609366985

Song, M. (2022). Meta's metaverse platform design in the pre-launch and ignition life stage. *International Journal of Internet, Broadcasting and Communication, 14*(4), 121–131. https://doi.org/10.7236/IJIBC.2022.14.4.121

Spielberg, S. (2018). *Ready player one.* Warner Bros.

Tambini, D. (1998). Civic networking and universal right to connectivity: Bologna. In R. Tsagarousianou, D. Tambini, & C. Bryan (Eds.), *Cyberdemocracy: Technology, cities and civic networks* (pp. 84–109). Routledge.

Turing, A. M. (1950). Computing machinery and intelligence. *Mind, 59*(236), 433–460.

Turkle, S. (1995). *Life on the screen: Identity in the age of the internet.* Simon & Schuster.

Turkle, S. (2005). *The second self: Computers and the human spirit.* MIT Press.

Turkle, S. (2011). *Alone together.* Basic Books.

UNESCO (2023) *DDS: De Digitale Stad/the digital city.* Retrieved March 15, 2025, from https://www.unesco.org/en/memory-world/dds-de-digitale-stad/digital-city)

Virilio, P. (2000) *The information bomb.* Verso Books.

Wachowski L., & Wachowski L. (1999). *The matrix.* Warner Bros. Pictures.

Weir, P. (1998). *The Truman show.* Paramount Pictures.

Žižek, S. (2000). From history and class consciousness to the dialectic of enlightenment... and back. *New German Critique, 81*, 107–123. https://doi.org/10.2307/488548

Zuckerberg, M. (2012, February 1). Mark Zuckerberg's letter to investors: "The Hacker Way". *Wired.* https://www.wired.com/2012/02/zuck-letter/

# Chapter 3
# Metaverse Spaces

**Abstract** This chapter examines how metaversal technologies are involved in the reconfiguration of spatiality, particularly on the level of virtualization and datafication. We first discuss how 'tech' companies conceptualize space in relation to their own business interests, generally by setting up virtual worlds and gaming platforms that draw in various social activities as well as give rise to new forms of labour and monetization. Next, we show how virtualization processes, such as the employment of game engines and the creation of digital twins, are positioning 'tech' companies as a central infrastructure across various industries. Last, we discuss how wearable VR and AR devices are used to capture physical spaces through virtualization and datafication, and what the implications of this are for the development of emergent AI systems.

## 3.1 A 'Visit' to Puerto Rico

A dominant yet relatively loose definition of the metaverse is that of a persistent, immersive, 3D environment in which people represented by virtual avatars interact with each other and other virtual elements or actors as well as carry out a broad range of professional and leisure activities. Needless to say, one of the primary sources of this definition is Meta. The metaverse is presented as "an embodied internet where you're *in* the experience" (emphasis added), which strongly suggests the confluence or collapse of boundaries between physical embodiment and virtual space (Meta, 2021). In comparison to many technology companies, which either do not use the term 'metaverse' or at least appear less committed to defining and employing it, Meta has not only fully embraced the term and publicly declared itself a 'metaverse company' but has also placed VR technology at the centre of this endeavour. To explore how the metaverse and the companies driving this endeavour are involved in the creation of new virtual spaces as well as the virtualization and datafication of physical spaces, Meta is a good starting point for examining how metaverse space and VR are presented and conceptualized, particularly in light of its history as a company that has exploited data on a larger scale than most.

© The Author(s) 2025

C. Hesselbein and P. Bory, *Infrastructures of Reality: Metaverse Stories, Spaces, Bodies*, PoliMI SpringerBriefs, https://doi.org/10.1007/978-3-031-97167-9_3

In early October of 2017, Mark Zuckerberg, the founder, chairman and CEO of the company that was then still called Facebook, and Rachel Franklin, the head of the Facebook's social VR division, went on a 'travel' to visit several 'places'. Somewhat surprisingly, their first stop was a 'visit' to Puerto Rico in the aftermath of hurricanes Irma and Maria. After a short stay, they 'teleported' onto the moon, and then they 'returned' to Zuckerberg's home in Palo Alto, California, to 'pat' his dog Beast.

As you might already suspect, Zuckerberg and Franklin did not physically travel to these places, but instead donned VR headsets and logged in to Facebook Spaces, the company's recently released platform for collective and immersive video watching. The entire event was broadcast live to Zuckerberg's approximately 90 million followers on Facebook. During their virtual visit to Puerto Rico, Zuckerberg and Franklin, or rather their virtual avatars, stood in front of, or in, a 360-degree video shot from the back of a truck driving through the flooded areas and documenting the destruction wrought by the hurricanes. While moving through this landscape, Zuckerberg and Franklin's rather cartoonish as well as legless avatars discussed how Facebook was aiding the relief efforts, such as donating money to the Red Cross and sharing mapping data. Perhaps equally importantly, while the video is showing the post-apocalyptic landscape of Puerto Rico, Zuckerberg also took the opportunity to demonstrate several technical features, such as the ability to 'turn in' and 'look around' the video as it is playing. Zuckerberg also emphasizes that "one of the things that's really magical about virtual reality is you can get the feeling you're really in a place". Then, to stress that they are "together in the same place and we're making eye contact", Zuckerberg and Franklin gave each other an enthusiastic yet awkward high five while appearing to stand in, or rather float above, a flooded intersection. Throughout the video, Zuckerberg appears to struggle to grasp the extent of the damage caused by the hurricanes or to remember their names, while grinning and adding vapid observations, such as "we are looking around" and "this street is really flooded". After repeating that "it feels like we are really here in Puerto Rico", Franklin adds that "it's crazy to feel like you're in the middle of it" before Zuckerberg suggests "teleporting" somewhere else and Franklin replies with a giggly "yeah maybe back to California".

Unsurprisingly, the livestream was widely mocked and criticized for being bizarre, tone deaf, and a form of disaster tourism that sought to promote Facebook's headset and VR platform on the back of the catastrophic death and destruction in Puerto Rico (Kastrenakes, 2017; Solon, 2017). The contrast between the VR-mediated voyeurism and the lived and ongoing suffering of the hurricanes' victims, vividly depicted against the backdrop of the conversation between Zuckerberg and Franklin, starkly brought out the harsh contrast between Puerto Ricans with no electricity or clean water and an American billionaire sitting in a comfortable office in the US. Despite using the term 'teleport' and avowing that they felt like they were 'really here' and 'in the middle of it', the VR space created with Meta's technology did not convincingly demonstrate that Zuckerberg and Franklin were genuinely *in* or *living* the experience of others. After the controversy and backlash in the media, Zuckerberg responded in a Facebook comment to the livestream:

When you're in VR yourself, the surroundings feel quite real. But that sense of empathy doesn't extend well to people watching you as a virtual character on a 2D screen. That's something we'll need to work on over time. [...] One of the most powerful features of VR is empathy. My goal here was to show how VR can raise awareness and help us see what's happening in different parts of the world. I also wanted to share the news of our partnership with the Red Cross to help with the recovery. Reading some of the comments, I realize this wasn't clear, and I'm sorry to anyone this offended.

Zuckerberg's reply reflects how VR technology has long been associated with a specific transformative potential to do 'good', even giving rise to what has been called the 'virtual reality for good' movement that seeks to employ VR for fixing a wide array of social issues, such as, among others, sexual harassment, racial discrimination, police violence, and even refugee crises and climate change (Carter & Egliston, 2024; Messeri, 2024). Indeed, VR has been labelled by some as the 'ultimate empathy machine' due to its purported capacity to transfer or build empathy in those who are exposed to VR. This idea has been widely taken up among filmmakers, entrepreneurs, technology startups and large companies, such as Meta. This connection between VR, virtual space, and empathy was popularized through a widely influential TED talk in 2015 by VR entrepreneur and filmmaker Chris Milk, who describes VR as follows:

"It's a machine, but inside of it, it feels like real life, it feels like truth. And you feel present in the world that you're inside and you feel present with the people that you're inside of it with". (Milk, 2015)

The conceptualization of VR as the 'ultimate empathy machine' is, on the one hand, heavily dependent on notions of embodiment and the transfer of affect between bodies (see next chapter). On the other hand, this conceptualization relies on a specific notion of presence and space. Through VR one can enter a new, immersive space to experience 'presence' and specifically the feeling of 'being elsewhere'. An underlying assumption that virtual presence in the same space with others will also signify a sense of social connection or care. This is why Zuckerberg and Franklin assert that they feel like they are 'really here in Puerto Rico' and Milk claims that it 'feels like real life'. The ostensible sense of reality and feeling of presence that VR can create have, however, been criticized for offering a false sense of proximity and empathy, because VR offers just an aesthetic or superficial mode of engagement with others and other places, thus reducing the victims and their painful reality to a mere, distant spectacle (Irom, 2021). Rather than real empathy, let alone real action, VR-mediated compassion engenders the illusion in viewers of having experienced authentic empathy without causing them to reflect on, for example, their own potential complicity or inaction (Nakamura, 2020). Zuckerberg's choice to present and demonstrate the abilities of Facebook's new VR platform against the background of a devastated landscape in Puerto Rico, similarly, emphasizes the privileged experience that he and Franklin are having through VR rather than the actual experiences of the victims still struggling in the aftermath of the hurricanes (Carter & Egliston, 2024). Rather than to "help us see what's happening in different parts of the world", Zuckerberg's use of VR and virtual space instead highlights the persistence of physical and emotional distance. In short, his experience was definitely not that of the flooded Puerto Ricans no matter how much they shared a virtual space. Moreover, Zuckerberg's response to the criticism not only situates the solution to the problem in the future ("work on over time"), which is a classical justification that 'Big Tech' companies have used to downplay the current harms of their actions, but he also re-emphasizes that VR is still the solution ("when you're in VR yourself"). That is to say, if only others could have been 'in VR' with him, they would be able to see and experience his empathy (Harley, 2020). In short, the real answer to such problems is that we need more VR as well as more people who are accustomed to it. The idea that virtual platforms or VR technology can create new and perhaps better places is nevertheless persistent and can be found in the narratives surrounding almost all companies involved in constructing the metaverse.

## 3.2  The Birth of Meta

Since Zuckerberg's unfortunate 'visit' to Puerto Rico in 2017, Facebook's ambitions in the fields of VR, AR, and virtual platforms as well as AI have only grown larger. Facebook's transition from a social media platform to a broader 'tech' company arguably started in 2014, when it acquired the VR startup Oculus, which had been founded just two years before as the result of a massively successful crowdfunding campaign on Kickstarter. In 2016, the first commercial, mass-consumer headsets were released (by Facebook as well as Sony and HTC), and in 2019 the company released the Quest, which was the first standalone (i.e. untethered) headset. By 2020, Facebook released the Quest 2, which found commercial success over the next two years, in part due to the lockdown measures introduced during the Covid-19 pandemic. However, it was Facebook's name change to Meta, and Zuckerberg's declaration of the pivot towards becoming a 'Metaverse company' in late 2021 that can be understood as the real kick-off of the metaverse hype. The announcement of the company's plan to build the metaverse by investing billions of dollars and hiring thousands of developers signalled a strong commitment to the idea that the future of digital technology lay in the development of VR. Soon after the announcement, Meta's Horizon Worlds platform was made publicly accessible, after an invite-only beta phase, to people in North America, and in the following years also to a larger number of countries in Europe and East Asia. However, some have suggested that the company's rebranding was likely driven by a need to cover up bad publicity (i.e. from the Cambridge Analytica scandal up to the release of the Facebook Papers by whistleblower Frances Haugen in early 2021, which exposed Meta's inability to tackle misinformation on its platforms as well as its failure to act on, among other things, online harassment). Another element in favour of Meta's rebranding was historical contingency. The Covid-19 pandemic had suddenly accelerated the acceptance of teleconferencing technologies, even by previously relatively marginal users who were excluded, or self-excluded, from such systems. The pandemic gave a strong push to the idea of working and socializing 'online only'; by late 2021 many people were already accustomed to physical isolation and online together-ness. In short, the time was ripe for Meta to push for a 'new virtual' world. Perhaps even more importantly, the rebranding appears driven by a crucial vulnerability in Meta's massively successful datafication business model, namely its dependence on the devices (smartphones and computers) and operating systems (iOS, Android, and Windows) of other large technology companies (e.g. Apple, Google, Microsoft) who therefore largely control Facebook's access to such data (Evans et al., 2022; Swisher, 2021). Indeed, Apple's introduction of the App Tracking Transparency (ATT) frame-work in 2021, which asks people to explicitly grant permission for apps to track their activity, severely limited Meta's ability to collect data for targeted advertisements. Meta does, however, dominate the market in VR hardware and operating systems, and shifting human activity in this direction would allow the company to maintain its highly profitable datafication practices. In other words, the metaverse, at least in Meta's version of this, would provide a new space for continuing commodification and data collection and for reaping their financial benefits.

What stands out about Meta's vision of the metaverse is that this new virtual space or 'world' will not only allow people to carry out a variety of unrelated activities—such as going to work, watching videos, playing games, socializing with friends, and exercising—but that this space will also encompass the various platforms in which such activities are currently being done. That is to say, Meta's Metaverse seeks to achieve a full-blown virtual world of networked platforms in which a vast range of hardware, software, and technical standards have been interconnected and made interoperable. Given Meta's history in terms of surveillance and the commercial-ization and exploitation of the data of those who employ the companies' multiple services, it is more than reasonable to suspect that Meta's version of the metaverse would continue along the lines of its previously established business model. Even by the standards of grand visions put forward by Silicon Valley CEOs and futur-ists, Zuckerberg's proposal is truly grandiose and all-encompassing. As even promi-nent metaverse boosters such as Matthew Ball and others have noted, creating such a massive, persistent, and interoperable network of real-time rendered 3D virtual worlds in one single framework is an enormously complex technological and organi-zation challenge. Although likely to be technically feasible, it remains unclear how near this potential future might be. It is precisely for this reason that scholars have suggested to focus instead on currently existing platforms or 'microverses' in which disparate activities such as working, playing, and consuming media are already being carried out, often with great popular success (Evans et al., 2022).

## 3.3  Gaming Platforms as Virtual Spaces

The experiences that VR technologies and platforms are purported to offer—of pres-ence, immersion, social connection, and empathy—and their discursive reiteration over the past decades form a powerful narrative around the new virtual spaces that are emerging on virtual platforms, even when these are not explicitly oriented towards either VR or AR. It is not for nothing that almost all if not all such platforms employ the concept of 'experience' to refer to the activities taking place there. Recall, for example, how Meta underscores that the metaverse involves being 'in the experience'. Indeed, the term 'experience' is central to almost all descriptions by technology companies of what takes place in virtual spaces. For instance, the large gaming and game-creation platform Roblox, which hosts millions of user-created games, exclusively refers to such games as 'experiences'. In the guidelines for its Creator Hub, moreover, Roblox even states that "a new place creates a new experience", thus entirely collapsing the distinction between the two (Roblox, n.d.). Experience is not just something one potentially has when one is in a place, but the place itself is an experience per se. Considering that many of these platforms, including Meta's Horizon Worlds, primarily focus on games, one might easily believe the claim that something more actively experiential is happening in virtual platforms compared to, say, scrolling down an Instagram feed or YouTube homepage.

It appears sometimes tempting to think of video games as a distinct and somewhat marginal or frivolous aspect of both current societies as well as scholarly research in media, communication, and technology. However, little could be farther from the truth. About 40% of the global population plays video games (of varying types, across various devices, such as computers, consoles, and smartphones), and this number rises to approximately 65% in rich countries. The value in terms of consumer spending has already outstripped that of cinema, music, and streaming video combined and is likely to soon overtake the amount spent on pay-tv (The Economist 2023). Gaming, in short, is big business. And almost all of the largest technology firms are heavily invested in gaming. For example, Microsoft, which launched the popular Xbox console in 2001, has taken over the game developer Activision Blizzard in one of the largest-ever technology deals in 2023. Apple and Google are central to game distribution because they own and operate the two main app stores, of which over half of their sales are of games. Amazon, through its Twitch platform, is the largest player, if you will, in terms of game streaming, an increasingly important form of media consumption. Similarly, in academic research, games and the field of games studies are no longer considered interesting but nevertheless somewhat marginal endeavours. Instead, games are now acknowledged as a "central component in the convergence of media content, media platforms and technologies, and media audiences" (Chess & Consalvo, 2022, 159).

Gaming platforms are a prime example of this convergence between previously disparate forms of media and practices that appear increasingly successful at drawing people into their virtual spaces to carry out various tasks and pastimes. Moreover, at least in terms of current and ongoing developments, the modes of production and consumption on such platforms appear to be of a different nature and open to the creation of new and different business models compared to social media platforms. For example, some user-generated games on Roblox have achieved large-scale success, attracting tens if not hundreds of thousands of concurrent players, thus spawning new business models and monetization opportunities. Equally if not more importantly, large gaming platforms are increasingly diversifying their offerings. They are becoming places not just for playing games but also for carrying out other activities, thus suggesting a role that is metaversal in the sense that they are succeeding at drawing in previously disparate activities into the same virtual space (Evans et al., 2022). That is, gaming platforms are positioning themselves as new spaces for experiences, practices, and pastimes that generally used to take place outside of virtual platforms.

The large gaming platform Fortnite, launched in 2017 by the US game and game-engine developer Epic Games, presents one good example of how gaming platforms are positioning themselves within broader socio-cultural structures and activities. Needless to say, Epic Games, the company behind Fortnite, and its CEO, Tim Sweeney, have been explicit about their ambitions to "build the metaverse and support its continued growth" (Takahashi 2024; Webster, 2022). The platform has multiple game modes. Fortnite Battle Royal and Fortnite Save the World are relatively straightforward (and enormously popular) shooter games. Fortnite Creative is a so-called sandbox game mode in which players can construct their own game maps, called 'islands', thus facilitating the creation of user-generated content. Also offered are Rocket Racing (self-explanatory) and Fortnite Festival, which is a rhythm game in which players perform songs using various instruments. In addition, there are two Lego Fortnite modes, developed in collaboration with The Lego Group, which reproduce the classical physical construction game, with several important modifications, in a virtual world. Compared to the main Fortnite modes, the Lego modes are family friendly, and notably similar, both in their contents and aesthetics, to another popular game platform that is frequently labelled as 'metaversal', namely Minecraft, which has led to suggestions that Lego Fortnite is mostly an attempt at competing with another metaverse rival (Cieslak & Gerken, 2023; Webster, 2024). Interestingly, Fortnite's video-game version of Lego has also made its way back into the physical form, that is, the classic plastic building blocks can also be purchased in a Fortnite-branded version, thus illustrating how virtual renditions and practices are bleeding back into the heretofore non-virtual world. All Fortnite modes are free-to-play except Save the World, which is pay-to-play, and all modes are compatible across different gaming platforms (i.e. players using different hardware can play together). In contrast to what one might expect, VR devices are not, however, supported by Fortnite due to the high levels of movement in their most popular games and the technical complexities of accommodating such locomotion mechanics without causing motion sickness in people wearing VR headsets (Stockdale, 2024).

The combination or mashing together of different brands and genres from multiple areas of pop culture is a fundamental characteristic of Fortnite, as well as other gaming platforms, which has collaborated with countless brands, artists, athletes, and influencers such as, among many others, Marvel, DC comics, Disney, and Nike as well as Travis Scott, Eminem, J Balvin, The Weeknd, and Lady Gaga. Such collaborations tend to take the form of in-game items, modes, or abilities, such as theme-specific challenges, featured 'skins' (outfits for avatars) or 'wraps' (to change the appearance of items, such as guns), limited-time modes (brief exceptions to the standard rules of a map), and custom emotes (generally short dances or poses). Such items and abilities are available for purchase in the Fortnite shop where in-game purchases can be made using the platform's currency called V-Bucks, which can be purchased with fiat currencies.

In addition to these multiple and highly successful game modes, Fortnite is also becoming a place for other activities that generally take place on other platforms or environments. For example, Fortnite has organized in-game concerts involving major musical artists and pop stars such as Travis Scott, BTS, and Ariana Grande, which attracted millions of attendees as well as many more who watched these events via livestreams on YouTube and Twitch. Moreover, in Fortnite's Party Royale mode, which is usually used to host concerts, movies are regularly streamed in a manner that looks similar to going to a drive-in cinema. In terms of other media, the Fortnite Shop allows you to buy music, again by major music artists, that can be performed in Fortnite Festival but also be played in-game. The Fortnite Shop, however, is generally a place where players buy virtual goods for their in-game characters or avatars, choosing from a plethora of items, accessories, and emotes that the platform has selected for this season's edition of their game modes. For example, one can purchase individual garments and items as well as whole outfits that carry the branding of, for example, popular TV shows such as Dragon Ball and athletes such as Shohei Ohtani, or more traditionally branded items, such as Nike Air Jordan sneakers. The shift from physical to virtual clothing as a mode of expressing one's identity is not just another indication of the ways in which gaming platforms are folding previously non-digital practices into their platform business models, but also an interesting development in terms of virtual embodiment and the expression of personal identity (see next chapter). Other smaller but nonetheless notable examples of the ways in which Fortnite is succeeding at enveloping new domains of activity, or at least becoming an additional space for such activities, are the JokeNite comedy club run by Trevor Noah, the multiple art exhibitions that have been organized on the platform, as well as museums, such as the Martin Luther King Museum and the Holocaust Museum, also known as Voices of the Forgotten (Webster, 2020; Wiener, 2024). Although such initiatives are probably relatively marginal in comparison to the numbers of players that come for more conventional gaming fun, they do indicate how gaming platforms are being considered as suitable venues for education, memorialization, and the presentation or preservation of creative arts and cultural heritage (Challenor, 2023).

Although Fortnite and Epic Games are excellent examples of the expansionary ambitions of gaming platforms and companies, perhaps an even more pertinent case is that of Roblox. Officially released in 2006, Roblox is often referred to as a game but is more accurately described as both a gaming platform hosting millions of different games as well as a game-development platform. In other words, you can create games for others to play or play games that you or others have created. In a certain sense, one could describe Roblox as a playground where one can pick between multiple games—or 'experiences', as the platform prefers to call them—that can be played either alone or with others.

There are many apparent similarities between Roblox and Fortnite as platforms, even if the games themselves, and particularly their gameplay and aesthetics, could hardly be more different. Like Fortnite, the platform is free-to-play, accessible via multiple devices, and has its own currency, Robux, which can be used for in-game purchases. Like Epic Games, the company behind Fortnite, Roblox is part owned by the large Chinese technology and gaming conglomerate, Tencent. Also similar is that Roblox has hosted concerts with well-known pop stars, such as Charli XCX, David Guetta, Lil Nas X, and Lizzo, and that brand collaborations are frequent. Spotify, for example, created an 'experience' in which players can create their own music and hang around in virtual music venues, Walmart set up an environment containing digital twins of its in-store products, which can subsequently be purchased through the company's e-commerce site, and Gucci created a virtual town as well as a garden, which were both temporary events lasting for a few weeks in which people could buy Gucci clothing and accessories for their Roblox avatars. Another similarity is that Roblox manages to attract millions of monthly players, slightly more even than Fortnite, with some games reaching hundreds of thousands concurrent players. Last, the founder and CEO of Roblox Corporation, David Baszucki, has indicated that Roblox is not only "building the metaverse" but also that the company and platform can be understood as its "shepherds" (Cross, 2021; Wakefield, 2021). In short, like Fortnite, Roblox can be understood as one particular platformized example of the metaverse in how it is succeeding at penetrating different economic sectors and spheres of social life as well as reshaping cultural practices and imaginations.

One major difference compared to Fortnite is that Roblox contains many more and more different games—an estimated 40 million games falling into various genres are currently being offered on the platform, with 5000 games attracting the lion's share of the approximately 200 million monthly players on the platform (Au, 2023). The most popular games on Roblox, such Brookhaven RP, Arise Crossover, and Adopt Me! frequently attract hundreds of thousands of concurrent players. Additionally, Roblox allows VR-enabled access to its platform, specifically through Meta's Quest headset. This enormous number of different games on the platform is the product of Roblox Studio, which is a so-called Integrated Design Environment in which anyone can create games that can be shared with and played by others on the Roblox platform. In other words, a marked contrast with Fortnite—although this platform does offer a Creator Portal through which people can also design their own game environments—is that user-generated games or 'content' are central to Roblox. Platforms that facilitated the creation of user-generated games are often referred to as 'sandboxes' because they focus on creative play rather than overt competition, are often open-ended, tend to lack a strong, centralized structure, and have close to limitless possibilities for creating a wide variety of things (Grimes, 2021).

Creating new games is frequently presented by Roblox as an opportunity for children—the largest user group on the platform—to learn coding and game design in a creative and engaging manner. Learning how to code through Roblox Studio is relatively easy yet challenging enough to have spawned an ancillary industry of tutorials with dedicated channels on YouTube, for example, attracting large numbers of subscribers and views. The creation of a Roblox 'experience' can be imagined as starting with a 3D blank canvas in which an environmental setting can be placed that can be subsequently filled up. Almost all details of this environment and its contents can be selected, although pre-made templates of various landscapes and stages exist, such as 'village' or 'western', as well as semi pre-made games, such as 'racing game' or 'obstacle course'. After this selection, one can start editing the terrain, placing objects and characters, design gameplay mechanics, script interactions, and add music and sound effects, among countless other things. In other words, the main and ostensibly most attractive feature of sandbox platforms such as Roblox is that they allow people to create and customize worlds according to their own wishes. These can eventually be shared with others to inhabit or play in, and, if found popular enough, can indeed become quite populous. If one prefers, however, one can also not choose to create games but instead focus one's attention on the creation of items, textures, animations, and audio, which can be sold to others via the Creator Store. The creation of items rather than games requires less time and effort, which is why more people tend to focus on the creation of such accessories rather than entire games. That said, most people on Roblox prefer to simply consume or play games rather than produce them, which is reminiscent of how a small number of 'prosumers' produced the majority of online content during the Web 2.0 era (Balbi & Magaudda, 2018). The vast majority of games as well as items available for sale through the Store are no longer created by Roblox but instead by people, often children, who invest considerable time and energy in creating them (Browning, 2020).

Roblox can be understood as one important way in which gaming platforms create new virtual spaces, not just in terms of the platform environment itself, but also in terms of the games that are created by people on the platform. Such user-generated and gamified spaces, in turn, can attract more players to the platform. Although there are millions of user-generated games on Roblox, the majority does not attract much attention. The existence of such 'empty spaces' or 'deserts' is not something new in digital media history. An early example of virtual wastelands are the myriad blogs and websites created during the age of the 'blogosphere', especially in the early 2000s, when users spent enormous amounts of time and effort to create their own webpages, which were often visited only by a few members or friends, and in some cases were never accessed except by their direct owners or creators. The mother of metaversal platforms, Second Life, has undergone a similar development; many abandoned spaces and digital ruins continue to mark its landscape (Veix, 2019).

Notwithstanding the desertification of some virtual places, the few thousand games that do succeed on Roblox, however, manage to bring millions of people into the platform at a time. The ability to create an almost infinite number of virtual spaces that are nested into the platform infrastructure suggests that the expansiveness of such spaces is limited only by the technologies that underpin them. That is, gaming platforms such as Roblox appear to facilitate a continuous expansion of virtual spaces or games that can be potentially populated with existing or new players. Moreover, both the platform as well as the games it hosts are managing to draw in activities and practices that were previously undertaken elsewhere as well as giving rise to new labour practices and business models. In short, they present one instance of metaversification in that they are drawing in previously-external industries and social activities as well as giving rise to new opportunities for commodification and monetization.

## 3.4  Virtual Labour, Playbour, and Monetization

One important sense in which Roblox can be also understood as a metaverse platform is that it is becoming a place for new types of content production and therefore also of new forms of labour and consumption. However, the content on Roblox is not simply that of established social media platforms, namely texts, images, and videos, but instead user-generated environments and games. Although both forms of content production—either on social media platforms or on metaverse platforms—can be understood as creative and therefore part of the so-called 'creator economy' (Bleier et al., 2024), both the possibilities of engaging with or consuming such content as well as the labour involved are qualitatively different. Whereas social media platforms consist mainly of communications between people, on sandbox platforms such as Roblox, people use the tools and services provided by such platforms to construct virtual spaces in which multiple and varied activities can take place. The labour involved in the production of this kind of 'content' is different compared to previous forms of content production, not least in the sense that one needs to acquire a basic sense of Roblox's coding language. As on social media, this labour can be pleasurable and rewarding, at times even in a financial sense, but it is the platform that ultimately benefits the most. However, the modes of consuming content are different as are their implications for social practices, such as the expression of identity through virtual apparel (see next chapter).

As a 'user' on a social media platform, engaging with user-generated content produced by, for example, influencers on YouTube or Instagram, involves consumption of the media, and potentially subscribing or following a channel or account as well as liking, sharing, and commenting on the content. The producer or 'creator' can monetize their content in various ways, such as, generally speaking, advertising to earn revenues per view, paid partnerships with brands, affiliate marketing to earn commissions on advertised products, and paid memberships or subscriptions. This system or business model in which content production, consumption, and digital labour are central is sometimes characterized as the 'influencer' or 'creator' economy (Duffy, 2017). On the platform level, all this content—including content that is not produced by influencers or with explicit commercial intent—is data that social media platforms collect and aggregate for various and well-established ends, such as business analytics, targeted advertising, data brokerage, and the building of AI models. As such, the creator economy can be understood as a part of the larger platform economy.

Engaging with content that is not merely text, audio, or video but instead a virtual world or game, such as happens on Roblox, enables the emergence of different monetization methods and business models. In this sense metaverse platforms are giving rise to new activities and possibilities for entrepreneurship as well as commodification and exploitation. And as will be discussed further below, monetization methods established on social media platforms will continue to apply, and advertising and datafication are highly likely to remain central if not further intensify (Hesselbein et al., 2024). However, gaming platforms such as Roblox allow for the creation of what appears to be a different and perhaps greater variety of content. Content on Roblox, for instance, consists not only of games but also of virtual objects, avatars, and emotes. That is to say, not just game environments are created by users but also the objects in such worlds as well as the avatars and the items involved in shaping their appearance. This suggests that content on Roblox is more actively and intensely engaged with than on social media platforms, particularly as 'engagement' involves more than watching and perhaps sharing or commenting but also playing, crafting, and interacting directly with other players. 'Metaversal' content, such as avatars and their appearances, are a part of players' visual identities as they move around the platform and between games, and therefore something that can be commodified and monetized both by content producers as well as the platform itself. Moreover, rather than indirect monetization of content through advertising, brand partnerships, and affiliate marketing, such as is common on social media platforms, content on gaming platforms and in the games that they host can be sold directly to consumers on the platform through the Creator Store or through in-game pay-to-access or pay-to-progress models. In other words, Roblox contains its own virtual economy of producers, intermediaries, and buyers of nearly all aspects of the broader platform, as well as its own virtual currency, Robux, which can be purchased with fiat currency. Through the Developer Exchange program, creators of games and in-game items are remunerated for their sales in the Creator Store, which can subsequently be exchanged back into fiat currency. As a result of the growth of Roblox and particularly the potential for financial reward, game development has become partially professionalized as game developers and game studios have started creating 'experiences' on the platform, particularly to develop 'advergames' for large brands as will be discussed in further detail below.

As mentioned, the majority of the Roblox economy is enabled through the labour of people, mostly children, and even though some are remunerated for this labour, the majority of the financial benefits, not to mention the potential value generated through advertising and the sale of data, go to Roblox Corporation. As it is mostly players who engage in the labour of producing both worlds and the economy of items, Roblox relies and capitalizes on 'playbour', that is, the confluence of play and labour performed by actors on the platform (Foxman, 2021, 2022; Kücklich, 2005). As a result, the platform has been accused of being involved in "industrial level commodification of children as free labour" (Evans et al., 2022, 48) What is more, since the population of labourers on Roblox consists mostly of children, this makes them not only vulnerable to labour or financial exploitation, but also other issues such as sexual harassment, hate speech, and scams. Roblox, as has been the case on social media platforms, has been slow, and frequently only partially successful at implementing content moderation, age restrictions, and other safeguards (Parkin, 2022).

Corporate advertising is another example of how Roblox is pulling in previously-external activities, which again illustrates how such gaming or metaverse platforms are becoming increasingly central to broader socio-economic infrastructures. Given the large amounts of time spent by many people on the platform, toy manufacturers and entertainment companies, among others, have recognized Roblox as an important place for advertising, especially considering the generally young age of players. Popular games such as *Adopt Me!* have seen the introduction of playable characters and items that are based on physical toys and brands, which subsequently—as seen earlier with the collaboration between Fortnite and Lego—result in physical toy lines based on Roblox games (Blackwood, 2023; Cucco, 2022). Unsurprisingly, the platform itself has sought to facilitate advertising in Roblox games through the introduction of digital billboards, posters, and other interactive surfaces, which game creators can integrate into their game designs and thus get a cut of the ad revenue (Sato, 2022). The turn towards advertising strongly suggests that Roblox has a greater incentive to collect data about the people on its platform in a manner that is similar to what social media platforms have done before, which can be used for targeted advertising. Putting up billboards and posters is, however, relatively conventional, to some extent mirroring existing practices in the physical world or in media production (e.g. product placement) as well as on social media platforms. The introduction of advertising into gaming—i.e. in-game advertising—is not necessarily new or specific to video games either, but when a video game is designed specifically around a product or brand it becomes a form of 'advergaming' (Tulleken, 2019). Such games are thought to capture one's attention more effectively because of their higher degree of interactivity, which makes this approach particularly suitable for children, who are more likely to not recognize the game's function as an advertisement as well as develop positive feelings for a given product or brand at a young age (Bains, 2024). It is therefore no surprise that a multitude of large corporations, ranging from clothing companies and fast-food chains to electronics companies and even other digital platforms (e.g. Netflix) have launched Roblox 'experiences' in recent years, thus signifying that the platform is moving beyond user-generated content and appears to be evolving in a manner similar to how social media platforms have done before.

One pertinent example of the entry of brands on Roblox and the practice of advergaming is Barbie DreamHouse Tycoon, which is a collaboration between game developer Gamefam and Mattel, the company producing the iconic doll. Created as a brand-driven virtual world (everything is pink and prominently features Mattel's name), the game involves creating and customizing a Barbie DreamHouse while gradually unlocking new rooms, decorations, and interactive features as players advance through the game. In addition to house building, players engage in role playing activities, such as dressing up their avatars, driving Barbie-themed cars, hosting parties, and interacting with branded items inspired by Barbie's cultural legacy. What is perhaps most striking is that Barbie DreamHouse is an advertisement that is inhabited, moved through, and played in rather than mostly audio-visually consumed. In other words, DreamHouse has a distinctly spatial and temporal dimension that is experienced in a more enveloping, long lasting, and captivating manner. To paraphrase Mark Zuckerberg, somewhat euphemistically, Barbie DreamHouse Tycoon is indeed an 'experience' in which one is inside. People accessing this branded space are not required to pay (i.e. freemium), but they are incentivized to purchase in-game content and services (i.e. pay-to-progress), including customization options, exclusive fashion items, home accessories, and vehicles, all available for purchase using Robux. Additionally, the game continuously implements limited-edition seasonal updates, featuring themed decorations and exclusive outfits tied to real-world seasonal trends. Such periodic 'drops' create a sense of urgency, encouraging players to make time-sensitive purchases (Cappa, 2025). In other words, the business model of such gamified spaces goes well beyond advertising but involves more active participation and indeed inhabitation by players.

Roblox, however, is not the only gaming platform that has been accused of exploiting playbour. Other platforms have also sought to newly create and draw in forms of labour onto their platform while benefiting from this activity in an asymmetrical manner. For example, Axie Infinity, one of larger blockchain-based gaming platforms, operates on a so-called 'play-to-earn' model in which players earn in-game currency—by breeding, raising, battling with, and trading characters called Axies in a manner similar to trading card games—that can be exchanged for various cryptocurrencies. Players need to purchase three Axies for a not-insignificant amount of money before they can start playing and earning the in-game cryptocurrency, which is why the model is perhaps more accurately described as 'pay-to-play-to-earn'. Moreover, each Axie character is a non-fungible token (NFT), which means its ownership is uniquely identifiable on a blockchain, one of the decentralized technologies that some maintain are central to the metaverse. Because players can earn money through playing—by exchanging the in-game currency for a cryptocurrency and eventually a fiat currency—Axie Infinity has found some popularity in countries with comparatively low wages. Although games with in-game currencies that are worth more than a local fiat currency have already acted as a source of income for people in countries with extremely devalued currencies, such as Venezuela (The Economist, 2019), this form of playbour became more attractive during the Covid-19 pandemic when platforms such as Axie Infinity became an alternative source of income in countries such as the Philippines where large numbers of people had become jobless (Chow & De Guzman, 2022). However, Axie Infinity is dependent on the continuous flow of new entrants who buy into the platform's currency, which has led to accusations of the platform running a pyramid scheme. Moreover, the relatively high entry costs have led to a system of exploitative arrangements between investors and labourers, and the high volatility of Axie Infinity's cryptocurrency, which is common to most such currencies, means that players can find that their earning have become close to worthless upon payout (Delic & Delfabbro, 2022; Ongweso Jr., 2022).

## 3.5  Virtual Property and Assetization

In addition to creating and exploiting new forms of labour, Axie Infinity also presents an example of assetization and financial speculation, thus asserting itself as a new virtual space for the continuation, intensification, and reconfiguration of existing economic and financial practices. Experiments with alternative forms of ownership, for example, are particularly common on blockchain-based platforms. Indeed, the

very idea of cryptocurrencies can be understood as a specific attempt at infrastructuralization in the sense that these seek to provide a decentralized alternative to the
conventional, centralized financial payment system. However, the extent to which
blockchains and cryptocurrencies are genuinely succeeding at providing an alternative to established financial infrastructures or whether they are fundamental to the
running of virtual economies on metaverse platforms very much remains to be seen.
In the latter case, it is already clear that the well-established virtual economies on
Roblox and Fortnite are operating without the employment of blockchains or cryptocurrencies, which suggests that this particular financial infrastructure is not, at least
for now, fundamental to metaverse economies.

Nevertheless, there are a number of platforms like Axie Infinity that continue
to assert, with varying degrees of success, the centrality of blockchains and cryptocurrencies, particularly through the purchase and trade of virtual 'land' and NFTs.
Moreover, like the larger and more successful gaming platforms, such blockchain-
based platforms have also sought to draw in various cultural activities, events, and
even government services. Decentraland, for example, has hosted several digital
fashion shows, Barbados has established a virtual embassy on the platform, and
companies such as J.P. Morgan Chase and Sotheby's have opened up virtual offices.
The Sandbox, similarly, has established numerous partnerships with major musi-
cians and celebrities, such as Snoop Dogg, Steve Aoki, Paris Hilton, and Gordon
Ramsey. Generally, such collaborations involve the creation of a specific structure
on the platform, where events and parties are hosted, games can be played, products
are sold, and NFT collections are released. Such NFTs are often playable characters on the platform, and can be purchased with the platform's cryptocurrency. The
extent to which such cross-over collaborations with other companies and brands are
successful is at best ambiguous. Both Decentraland and The Sandbox appear to have
relatively low numbers of active visitors and often feel deserted, as becomes readily
apparent when visiting them (Au, 2023).

Both Decentraland and The Sandbox, although self-described gaming platforms,
can be generally characterized as blockchain-based attempts at seeking to reproduce
in virtual spaces the idea of ownership over land and objects. The replication of real
estate markets in virtual spaces has attracted considerable attention in recent years,
not least because large amounts were invested in plots of virtual space, which in some
cases were sold for millions of dollars. Major companies, brands, and celebrities,
ranging from Samsung and Adidas to Tommy Hilfiger and Heineken, invested in
property on Decentraland, often setting up digital stores in such gamified spaces as
well. Interestingly, some scholars have argued that virtual real estate markets "are not
simply a 'feature' of those virtual worlds, but rather one of the main drivers of their
development" (Baumann & Fauveaud, 2024, 3), thus again underscoring how virtual
platforms can be understood as attempts at establishing new spaces for the drawing
in of existing social and economic activities as well as creating new opportunities
for capital accumulation (Alvarez León & Rosen, 2024). NFTs can similarly be
understood as not only a process through which virtual objects can become assets—
by securely establishing their ownership on blockchains and creating scarcity—
but also a means for opening up physical objects to new forms of monetization

and financial speculation in virtual spaces (Belk et al., 2022; Wiener, 2022). The hype around NFTs peaked in 2021–2022—most notably with the sale of Beeple's *Everydays: The First 5000 Days*, a collage of thousands of digital images, for close to 70 million dollars—and has since somewhat faded from popular discourse due to the extreme volatility of cryptocurrencies and a number of prominent scams and scandals. Nevertheless, an active community of sellers and traders continues to exist, not least on virtual platforms such as Decentraland and The Sandbox. This cryptocurrency activity perhaps explains the continuing existence of blockchain-based platforms, which otherwise tend to feel largely deserted. Indeed, as gaming platforms they appear not amount to much, but as investment and speculation vehicles they continue to exist. To paraphrase Evgeny Morozov, Decentraland and The Sandbox are perhaps not so much in search of a territory as they are of a population (Morozov, 2022).

Taken together, phenomena such as virtual real estate markets as well as NFTs can be understood as part of a broader move towards assetization and new forms of 'rentier capitalism' that characterizes the digital platform economy (Birch, 2020; Sadowski, 2020). Unsurprisingly, games and gaming platforms similarly reflect this shift in terms of ownership and assetization. Fortnite and Roblox, for example, do not require gamers to buy the games or even a console, but instead encourage them to pay for seasonal 'battle passes' or subscription fees as alternative forms of commodification and monetization (Joseph, 2021). Moreover, the reconfiguration of the logic of ownership is also reflected in such games themselves, namely in the form of almost infinitely customizable attires and accessories for avatars in addition to the plethora of items (e.g. weapons, tools) and actions (e.g. emotes, poses) that can be bought on such platforms. In other words, Fortnite and Roblox function as highly profitable assets because of their ability to continuously draw income, or 'rent', from players through the commodification of in-game items while simultaneously using player data to further optimize and personalize the games for further financial extraction (Bernevega & Gekker, 2021).

## 3.6   Datafication of Virtual Spaces

The datafication practices of digital platforms and the role such data have played in, on the one hand, the commodification and exploitation of people and their behaviours and thus surveillance capitalism (Zuboff, 2019), and on the other, the development of recommendation algorithms and AI systems, are well documented. The role of data on gaming practices is, however, relatively underexplored (Seif El-Nasr et al., 2013). An emerging body of scholarship notes that, in the shift of digital platforms towards assetization, income is generated from such assets in two manners, namely in the form of money and in the form of data (Sadowski, 2020). As already noted, money exchange is generally eschewed in favour of the value that can be generated from data collection, which is why search engines and social media platforms appear 'free' to use. In other words, companies offering such 'free' services have to extract data and generate value from them in order to continue to operate. Similarly, a crucial development on gaming platforms has been that monetary exchange is often foregone (free-to-play or freemium) in exchange for so-called 'data rents', which are close to impossible to avoid once one enters a gaming platform (Bernevega & Gekker, 2021).

Data from gamers—game analytics—is used to track, among other things, in-game actions and behaviours, interactions with other gamers and the virtual environment, time spent playing, in-game purchases, as well as their personal identifiable information (e.g. name, location, age, gender), and device details (e.g. hardware specifications and software versions) (Kerr, 2017; Seif El-Nasr et al. 2013). Moreover, although data from gamers can be directly sold, there are several important additional forms of value that can be extracted from this. For instance, data can be used for profiling and targeting, optimizing and improving technical systems or organizational structures, generating novel business models, creating new products and services, and increasing the value of existing assets (Sadowski, 2019). In the context of gaming, data can therefore be used to improve game development, tailor the game to individual players, and to suggest further gaming content, but data also serves to target ads as well as be sold on to data brokers. Moreover, such game analytics can be turned into game metrics to quantitatively represent a game's performance and economic success as well as the behavioural patterns of gamers—particularly to investors or shareholders—thus increasing the value of the asset at hand, namely the gaming platform. Indeed, as Bernevega and Dekker note, "games are not only designed in accordance with insights derived from data but also designed to extract more data" (2021, 63). According to current scholarly literature, it is still relatively uncertain precisely to what end games and game media can be datafied or put to use in terms of the development of, for example, AI systems. However, this was also the case for the pictures and videos that were being uploaded to social media platforms in the mid-2000s, when people doubted whether anyone would be interested in such personal content or to what ends such media could be put to use. Indeed, many forms of data appear to be collected mostly for their potential future use and profitability. Moreover, interactions between large numbers of people in virtual worlds are considered to be a potentially rich source of information about social networks and their evolution (Shah & Sukthankar, 2015) as well as marketing practices, consumer transactions, and social trends (Ahmmad et al., 2023). What is clear, however, is that the virtual spaces created on gaming platforms, such as Fortnite and Roblox, are so far very successful at extracting regular monetary value from gamers via in-game transactions and subscriptions while simultaneously collecting perhaps an even more valuable commodity, namely their behavioural data.

## 3.7    Virtualization of Infrastructure and Datafication of Space

Roblox and Fortnite are important examples of the ways in which games are becoming metaverse platforms or 'microverses' by providing a virtual space in which multiple actors can interact with each other as well as drawing in and datafying practices that used to be done elsewhere. Although Fortnite is an important and thus far apparently successful form of the platformization of gaming and of culture more broadly (Nieborg & Poell, 2018), equally important is Epic Games's success at finding broader uses for one of its core technologies, namely the game engine. In Epic's case, the game engine at hand is called Unreal Engine, which is a 3D creation tool that was originally developed for the first-person shooter game Unreal, which was released in 1998 and is one of the most successful videogames in history. In a nutshell, game engines are a means of creating virtual spaces as well as the structures, items, persons, and phenomena (e.g. light and sound) found therein as well as the ways in which such diverse phenomena can interact (game physics). Together with Unity, another widely employed game engine, Unreal has been used for a number of highly successful games, including AR and VR productions such as Pokémon Go and Beat Saber. Unreal Engine as well as its popular competitor Unity are fundamental to all aspects of the production of games and together have a dominant market share. Although such game engines can make game production accessible to more than just professional game developers, they appear to increasingly position themselves as a platform monopoly—i.e. seeking to dominate the game development market by setting the standards of interoperability, crowding out competitors, making users dependent on their tools and services, such as 3D asset stores, and expanding their influence into new markets (Foxman, 2019).

Game engines have, over the past 30 years or so, become capable of rendering close to photorealistic graphics and recreating 'real-world' physics. As a consequence, their abilities have become of use in other industries that similarly seek to recreate and represent new or imaginary worlds, such as cinema and film production where so-called virtual sets have become central in the making of recent and highly successful films and televisions series ranging from *Barbie* and *Megalopolis* to *The Mandalorian* (Freedman, 2020). Virtual sets can be understood as studios in which, rather than a traditional, physical backdrop (or a green screen), large LED screens are used on which computer-generated images can be displayed (and manipulated) in real time. In other professional fields that frequently need to visualize 3D designs or plans, such as architecture, design, engineering, and construction, game engines have also become increasingly important for the modelling of buildings and large civil engineering projects (Chia, 2022; Chia et al., 2023; Jungherr & Schlarb 2022). In addition to drawing more and more activities and practices into platforms, metaversal technologies such as game engines—through their facilitation of core functions in diverse industries—are increasingly taking on the form of an infrastructure that goes far beyond the realm of video games.

The transition of game engines from a technology central to the development of computer games to other cultural industries involved in the production of audio-visual media as well as professions that frequently rely on visualization is not unexpected. However, digitalization and virtualization are also at the centre of the emergence, over the past two decades, of the so-called digital twin. Importantly, this is an instance in which technology is used to virtualize existing physical spaces and processes, thus providing one example of the expansionary tendencies of metaverse companies towards capturing and datafying the external world. In contrast to the virtual spaces and gaming platforms described above, which seek to draw existing practices into their ambit, here we see how the logics of virtuality are pushed out into the existing world. As a consequence, the structures, objects, and processes captured through such virtualization technologies are increasingly opened up to control by the companies engaged in such endeavours. Unsurprisingly, game engines and gaming companies play an important role in this process because of their expertise in virtually replicating physical objects and dynamics as well as rendering them in virtual environments. Equally unsurprising is that microchip manufacturers such as Nvidia, which initially emerged as a producer of Graphics Processing Units (GPUs) that are fundamental to the rendering of computer graphics and more recently to supercomputing and AI, are also heavily involved in the creation of digital twins.

In a nutshell, a digital twin is a replica, model, or simulation of a physical object, process, system, or even body. Generally speaking, such 'twins' are used to monitor and model, through real-time sensing and data collection, the characteristics, operations, and processes of the original, physical counterpart. Digital twins can be used to simulate, test, and experiment with changes to the structure or operation of a physical system, such as the integration of new features, without potentially hampering or endangering the functionality of that system. Such virtual experimentation through adaptive models can be used to identify opportunities for implementing changes and optimization, but also to predict likely failures and mitigate unexpected and adverse developments. Although initially a form of computer simulation in which information and control between virtual renditions of physical phenomena were not directly interconnected in real time, digital or virtual twins are used to not just model physical phenomena but also to directly intervene in the operations of a physical structure or system. Moreover, because of the complex nature of aggregating and integrating diverse types of data (i.e., multi-modal data) collected by various sensors, AI systems capable of performing such operations have become increasingly central (Kreuzer et al., 2024). Due to their ability to monitor and manage complex systems, digital twins have found uptake in a number of settings, such as large infrastructures (stadiums, airports, factories, cities) and complex technologies (aircraft, wind turbines, oil rigs, and space rockets), and processes that extend across geographical spaces and multiple actors, such as supply chains in the manufacturing industries (Jebelli et al., 2024). In addition to such industrial applications, digital twins are also being employed in the tourism and heritage sectors, where they can be used to provide virtual alternatives to overly popular tourist attractions or for the recording and preservation of heritage sites that are at risk of disappearing. Most dramatically, some countries under the threat of climate change, such as the island state of Tuvalu, have announced the wish to digitally preserve a digital twin of their nation and culture in metaversal environments (Yeo, 2024).

The emergence of digital twins and the convergence of 3D modelling, data collection via ubiquitous sensors, and AI processing is referred to as the 'industrial metaverse' (Bohné et al. 2023). A particularly pertinent example of the relatively recent convergence of such previously disparate technologies and processes can be found in Nvidia's Omniverse platform. Although Nvidia is best known as the world's largest microchip maker by market capitalization as well as the company whose chips have been foundational for development in supercomputing, machine learning and AI over the past decades, it certainly also claims to have a stake in the metaverse. The CEO of Nvidia, Jensen Huang, for example, has asserted that the metaverse "is where we will create the future" (Shapiro, 2021) and that "the economy of the virtual world will be much, much bigger than the economy of the physical world" (Hyman, 2021). A closer look at how Nvidia envisions the metaverse—and particularly its relationship between AI systems and the development of digital twins that virtualize physical infrastructures and processes across multiple industries—is therefore merited.

Nvidia Omniverse can best be described as a collaboration platform for creating and running 3D applications and services, such as, among others, rendering graphics for design projects, creating visual effects for video productions, and simulating the behaviour of entities in 3D environments. The combination of software applications, (AI) tools, graphics engines, datasets, and computing services are, in other words, useful for professionals in various sectors of the entertainment, architecture, and design industries as well as engineering projects in fields such as transportation and construction. What differentiates Omniverse from game engines such as Unreal and Unity is that it seeks to bring together in one place a 'suite' of different tools as well as types of data that are usually challenging to work with collectively and simultaneously. In short, it allows one to work with multiple graphics renderers, design tools, image editors, and game engines at the same time. What makes Nvidia Omniverse perhaps most 'omniversal' is that projects can be visualized in real time and in the same digital setting using the Universal Scene Description (USD) format—an open-source framework for the exchange of 3D computer graphics data—which is already widely used in media and design industries. Importantly, Nvidia, together with other members of the Alliance for OpenUSD (e.g. Pixar, Adobe, and Apple), seeks to position OpenUSD as a global infrastructure for the standardization and development of 3D content much in the way that HTML did for the internet (Cherney, 2021). In short, OpenUSD seeks to become an infrastructure that undergirds graphics production. Needless to say, the computation and rendering of virtual objects and environments as well as the physics of such environments are powered by Nvidia's own microchips.

Nvidia asserts that its Omniverse platform "plays a foundational role in the building of the metaverse, the next stage of the internet" (Nvidia n.d.). The metaverse as imagined by Nvidia is, however, in marked contrast to that of other companies which focus more on drawing in practices of leisure, consumption, and socializing. Instead, for Nvidia, the ambition appears to become a key structural node at the centre of multiple modes of industrial production by positioning itself as an 'obligatory passage point' not just in terms of its microchips but also through its Omniverse work environment and toolset (Smith, 2024). Moreover, the platform's integration of AI and machine-learning systems allows it to collect, analyse, and provide insights based on the operational data generated through usage of its suite of applications. One element of this is the development of digital twins, which are increasingly becoming mediators between virtual modes of interface and control with physical structures and processes. Such applications are currently being used in the operations of large companies with complex supply chain and manufacturing processes, such as Amazon, Siemens, and BMW, which are increasingly seeking to introduce robots into their warehouses and factories. Digital twins allow for extensive simulation and training of the AI systems employed not only in the management of the data flows created by such environments but also of the robotics introduced into them. However, what is perhaps most interesting about Nvidia Omniverse is that its collection of functionalities in terms of virtualization is being applied in large infrastructure projects. Germany's national railway operator Deutsche Bahn, for instance, is creating a country-scale digital twin to fully simulate operations across the entire German train network (Geyer, 2022), thus suggesting in at least one way how virtualization technologies, and therefore metaverse companies such as Nvidia, can become a potential 'infrastructure of infrastructures'.

In addition to the virtualization of built environments through the development of digital twins, another important means of pushing metaverse logics into the physical and social world, or perhaps more accurately, blending the virtual with the physical, is that of mixed-reality wearables, such as head-mounted VR displays and AR 'smart' glasses. One primary means of doing so is how people wearing such technologies quite literally bring the virtual into the physical world, often, but certainly not necessarily always, in a highly visible manner. A person wearing an AR headset in public, such as the Apple Vision Pro, is highly likely to attract quite a bit of attention. Someone wearing Ray-Ban Meta glasses, however, is much less likely to be noticed. Nevertheless, both technologies, by their mere presence in public or domestic space, open up such spaces to important forms of datafication by the companies that produce these technologies, most prominently Meta. In doing so, wearables extend the logics of the metaverse into our world, and by consequence also the reach of the companies that expound this specific techno-utopian vision. Moreover, the dominant imaginary surrounding VR and AR has long been that of gaming devices, but the producers of such devices are increasingly asserting their everyday use as means of communication and social interaction (Egliston & Carter, 2020). If and when VR and AR in fact do become ubiquitous—moving through various spaces, applied in various contexts, employed in multiple professional and private practices—their value as datafication devices becomes all the more important, which has important implications for how we experience such virtualized spaces.

A first example of corporate attempts at reconfiguring spatiality can be found in the promotional discourse surrounding the introduction of Apple's headset, the Vision Pro. Notably, Apple is one of few companies that does not seek to affiliate itself with the term 'metaverse'. Instead, it tends to refer to 'augmented reality'.

Perhaps tellingly, the Vision Pro headset is billed as a 'spatial computer', which indicates Apple's attempt at setting boundaries between its own conceptualization of this technology and that of other companies, thus strategically positioning itself as unique. In practice, however, the Vision Pro is best described as a mixed-reality device that can be used for both VR and AR applications. That is, when wearing the headset, one can either completely immerse oneself in a virtual reality environment or add an augmented reality layer to one's perceptual field.

Apple is a company that has succeeded better than almost any other competitor at shaping 'tech' narratives and establishing an almost cult-like level of devotion among consumers, most notably through its founder, CEO, and prime mythmaker, Steve Jobs (Magaudda, 2015). Moreover, as the company that is perceived as 'disrupting' not just personal and mobile computing but also the music, film, and publishing industries, Apple's entry into the field mixed-reality devices is therefore not to be underestimated. Although the Vision Pro might well turn out to be a commercial failure—the costs of the headset are considerable even without peripheral devices and add-ons, but this was also true for the first iPhone models—Apple, as one of the largest technology companies, has thrown its considerable resources behind the development of a 'metaversal' device and is seeking to shift the discourse around this technology more broadly. One of the prime examples of this is Apple's imaginary of spatiality—not least brought out by its emphasis on 'spatial computing'—and particularly how company discourse around the Vision Pro seek to position the headset (and other Apple products) as having a normal and desirable place in everyday life. Research on the promotional materials used during the introduction and release of the Vision Pro, for example, has noted how Apple conceptualizes the headset's usage and its users (blending personal and professional life in domestic and public settings), its relationship to physical environments (continuously captured and remediated), to Apple broader 'ecosystem' of products (seamless integration with other Apple devices), and in relation to spatiality in general (bringing digital applications 'in your space' (Blackman & Harley, 2024). As one might expect, privacy concerns and datafication practices, as much as infrastructural needs and environmental impacts, are largely absent from materials communicating Apple's vision of 'spatial computing', generally presenting the headset's technological operations as natural and desirable.

As previous research has noted (Egliston & Carter, 2021a, 2021b), however, a primary concern with mixed-reality headsets and other wearables is the collection of data about built environments that people move through as well as work and live in while wearing technologies such as the Vision Pro. Spatial understanding is not only necessary for mixed-reality technologies in order to combine physical environments and people with (partially) virtual elements, but, particularly in the case of VR, to safeguard users against potential external harms. Such abilities allow for the collection of spatial data on, for example, geolocation, the dimensions and layout of one's personal or workspaces, the presence and placement of various type of furniture or devices, and the state and quality of various objects in such spaces (Egliston & Carter, 2022; Nair et al., 2023). Needless to say, such information can be used to infer one's financial capabilities and aesthetic preferences and is thus hugely valuable for advertising as well as other purposes.

Although Apple is a powerful shaper of 'tech' discourse, it is Meta that is the most prominent company in terms of shaping not just the discourse around VR/AR but also the production and dissemination of such headsets. As already noted, Meta (then still Facebook) first sojourned into VR with the acquisition of the startup Oculus in 2014 and has since become the dominant force in the wider VR market with the release of several commercially successful models of its Quest headset. Although Apple is a company that has a relatively good reputation in terms of privacy and data collection—the Vision Pro claims not to send its data to third-party apps, for example (Blackman & Harley, 2024)—Meta is a company that definitely does not. Above all, this is the result of numerous incidents and scandals involving the Facebook platform (e.g. Cambridge Analytica) as well as the company's reputation in terms of failing to safeguard the data of its users as well as exploiting such data in numerous ways. Privacy concerns also extend to the level of Meta's devices. Research, for example, on the privacy and data policies of Facebook's first iterations of its VR headset, the Oculus, discursively frame the company's data extraction processes as responsive to and responsible in debates about data privacy (Egliston & Carter, 2021b). In other words, such privacy and data policies position the company within a favourable narrative in which it seems to adhere to regulations and best practices while also appeasing potential concerns from users, regulators, and researchers. Similar critical research into Meta's hardware, such as the Oculus VR headset—as an assemblage of sensors, cameras, and computational processing—has underlined its data intensive capabilities not just in terms of the collection of data about bodies but particularly of environments, namely by enabling "the device to algorithmically construct a map of the environment around itself, such that the software can track the position and movement of the device through space" (Egliston & Carter, 2021a, n.p.). In short, spatial information, of the environment but also of users as they navigate spaces, is one of Oculus's main data products.

Meta's interest in tracking, virtualizing, and datafying the environment is also apparent from its acquisition of startups specializing in the mapping and computational modelling of physical space. Since 2019, the company has acquired Mapillary, a company that crowdsources maps based on image data, Scape Technologies, a developer of 3D mapping and scene reconstruction processes using computer vision, and GrokStyle, a company centred on mapping and virtualizing retail spaces using AR (Carter & Egliston, 2024). A series of research and development projects inside Meta betray a similar interest in integrating mixed-reality technologies into the ways in which people move through the world. Livemaps, for example, a now defunct project was presented as a 3D 'shared virtual map' that relied on crowdsourced information from 'smart devices' that could be navigated with the use of AR headsets and 'smart glasses' (Reality Labs, 2019). In a similar vein, more recent projects such as Aria and Ego4D seek to combine computer vision and AI processing by collecting data in the form of scans and videos of the environment—from an 'egocentric' or first-person perspective—for the further development of VR and AR headsets as well as AI. VR technologies have long been central to Mark Zuckerberg's vision of the metaverse, but AR appears recently to be taking on a more prominent position within this vision. Another interesting recent development is a shift in language about AR, which appears to be moving from 'headsets' towards 'glasses', interestingly suggesting a closer relationship to the 'smart glasses' recently released in a collaboration with Ray-Ban (Carter, 2024; Meta, 2025). If such glasses indeed become widespread, the potential of such devices to map and datafy the environment, both domestic and public, is enormous, thus giving rise to further concerns about datafication and privacy (McArthur, 2024).

Meta is certainly not the only company involved in mapping, tracking, and virtualizing the environment. In 2019, Epic Games acquired Quixel, a company that makes 3D models of objects based on real-world high-definition photography, a process also known as photogrammetry. The word 'objects' is perhaps a bit misleading as Quixel—in addition to digitizing myriad manufactured things and natural phenomena—also engages in so-called megascans of landscapes and even entire countries, such as Iceland (Wiener, 2024). Besides expertise in scanning and virtual rendering, Quixel also operates an enormous 'asset library' where digital designers and artists can buy, share, or sell virtual scans. Such scans, once integrated into Epic Games's Unreal Engine, make, it is claimed, 3D world building 'easy'.

Mixed-reality technologies, such as the Vision Pro and Quest, present examples of how our relationships between physical and virtual reality are currently developing into what some scholars have called 'coextensive space' or "an altered relationship between the physical, the digital and concrete space, through the mediated inclusion of concrete reality" (Saker & Frith, 2020). The collection of spatial data is not, however, limited to the inside spaces (either domestic or professional) where VR headsets, for now, still tend to be worn. As just noted, technology companies are also seeking to datafy public environments in ways that go beyond mere mapping, such as most famously done by Google. Here the grandiose ambitions underlying the development of the metaverse once more become apparent. Meta's aforementioned project Aria, for example, has been characterized as seeking to map and collect data on the 'Commons'—non-privately owned and publicly maintained spaces—through sensory devices such as mixed-reality headsets and glasses, which in turn will be further developed on the basis of such data (Applin & Flick, 2021). However, this represents an advanced form of 'data colonialism' (Couldry & Meijas, 2019; Thatcher et al., 2016) that goes beyond the textual, audial, and visual data found on social media platforms, this time of spatial and interactional data pertaining to public spaces, and once again, collected without asking consent or seeking public input on the desirability or necessity of such an endeavour. Taken together, such data on a wide range of spaces, spatial qualities, and spatial dynamics, brought into combination with data on people and their behaviours, can subsequently be marshalled for more intense forms of psychometric profiling and subsequently monetization, commodification, and targeted advertising.

## 3.8 Colonizing Space

Much of the hype around the metaverse, NFTs, and blockchains was replaced in late 2022 by the frenzy that emerged after OpenAI's release of the GPT-3.5 model and the ChatGPT interface. That frenzy has since reached a fever pitch, as OpenAI released further models, ChatGPT started to be used in various professions, and other companies and countries started releasing their own chatbots, image generators, and so on. As we know, the operation of such generative AI systems was and is dependent on vast amounts of energy and data, especially, text, image, and audiovisual materials that have been extracted and appropriated by large technology companies from the internet.

Large Language Models (LLMs)—the AI systems that provide the answers given by such chatbots and that underpin their ostensible 'intelligence'—produce impressive results in the well-delineated setting of text interfaces and linguistic interaction. Moreover, AI, and generative AI in particular, are seen as fundamental to creating much of the textual, audiovisual, and architectural content of virtual or metaversal environments, thus making them more realistic and immersive, as well as enabling 'natural' interactions with the non-human actors in such places (Lv, 2023). However, LLMs have a poor understanding of the natural and built environment because they are trained on text and image data that has been culled from the internet. This means they have only a limited grasp of the physical world and its various properties, and are therefore not very useful for integrating AI systems into robotics systems, which is considered by many to be one of the next frontiers in technological development. Once robots are capable of understanding, for example, the spatial properties in which they will work or interact with human and non-human animals, they will, it is thought, be safer to be around, more productive, and also more likely to be accepted by humans.

Prominent AI scientists, such as Fei Fei Li for example, have asserted that the next big step for AI is to "understand how the world works" (Li, 2024). One important element of this sense of understanding the world is, of course, spatiality and the physics of activities and interactions in space. Future AI systems, however, particularly robotic ones that will operate among and interact with humans as well as move in spaces where humans work and live, will, like LLMs, depend on large amounts of data. Data collected in virtual spaces as well as through the virtualization of space, one could call these metaversal data, are therefore likely to play an important role in the development of AI systems and AI-powered robots. Such models—which combine text, image, and video about more diverse phenomena from more varied contexts—are thought to be crucial for providing generative AI systems with a much richer, better sense of 'how the world works'.

In order to develop so-called multi-model AI systems that are capable of processing multiple types of data, it is necessary to collect multi-modal data, that is, data on more and more diverse phenomena as well as to create models that contain more information than data about language. As a result, a variety of large spatial models, large behavioural models, and large world models have recently been released, each seeking to fill this gap. For example, Niantic, best known for developing popular mobile AR games such as Pokémon Go, has released a 'large geospatial model', which purportedly enables AI systems and robots to "not only to perceive and understand physical spaces, but also to interact with them in new ways" (Brachmann & Prisacariu, 2024). Similarly, Nvidia has announced the development of 'physical AI' and the integration of such systems into its Omniverse platform, which is purportedly capable of understanding spatial relationships and physical behaviours of the multidimensional world. Such 3D training data can be collected from, in addition to the 'real' world, computer simulations conducted in digital twins, which can serve as both a source of data and a training ground for the development of AI systems. The generation of such synthetic data is an important part of the new frontier of datafication, not only because bigger, newer, more varied data are considered crucial for AI systems but also because such systems are believed to be running out of high-quality data to train on in the near future (Xu, 2022).

Although it is too early to indicate precisely how the relationship between developments in metaversal technologies, data, and AI will further evolve, it appears that the relationship is more one of symbiosis than replacement. In short, the hype surrounding the metaverse and the current furore around AI are deeply intertwined. Indeed, it is becoming increasingly clear that 'Big Tech' companies are seeking to actively integrate AI systems into robotics, specifically for modelling and processing of the built environment. Google DeepMind, for example, recently released two Gemini AI models that supposedly give robots "the humanlike ability to comprehend and react to the world around us" (Google DeepMind, 2025). And at its recent developer conference, Nvidia, in partnership with Disney, introduced a new robot called Blue, which is also capable, it is claimed, of modelling and processing its physical surroundings. It is entirely unsurprising that one of the first forays of such an AI-powered agent into our society, developed by two major technology and entertainment companies, would look like a seemingly disarming and lovable creature, inspired by Star Wars, a movie series that revolves around the colonization of space.

## References

Ahmmad, K., Howlett, E., & Perkins, A. (2023). Into the matrix: Collecting psychometric data from consumers immersed in virtual worlds. *Journal of Marketing Theory and Practice, 32*(3), 1–18. https://doi.org/10.1080/10696679.2023.2202861

Alvarez León, L. F., & Rosen, J. (2024). Land, reconfigured: Defying the laws of physics, upholding the rules of the market in the metaverse. *Environment and Planning D: Society and Space, 42*(4), 559–579. https://doi.org/10.1177/02637758241257115

Applin, S. A., & Flick, C. (2021). Facebook's project Aria indicates problems for responsible innovation when broadly deploying AR and other pervasive technology in the commons. *Journal of Responsible Technology, 5*, Article 100010. https://doi.org/10.1016/j.jrt.2021.100010

Au, W. J. (2023). *Making a Metaverse that matters: From snow crash & second life to a virtual world worth fighting for*. Wiley.

Bains, C. (2024, July 13). "Advergames": How games platform Roblox became a corporate marketing playground. *The Guardian.* https://www.theguardian.com/games/article/2024/jul/13/advergames-how-games-platform-roblox-became-a-corporate-marketing-playground

Balbi, G., & Magaudda, P. (2018). *A history of digital media: An intermedia and global perspective.* Routledge.

Baumann, Y., & Fauveaud, G. (2024). Metaverses and virtual real estate markets: The commodification and assetization of the digital. *Economy and Society, 53*(4), 557–578. https://doi.org/10.1080/03085147.2024.2408984

Belk, R., Humayun, M., & Brouard, M. (2022). Money, possessions, and ownership in the Metaverse: NFTs, cryptocurrencies, Web3 and Wild Markets. *Journal of Business Research, 153,* 198–205. https://doi.org/10.1016/j.jbusres.2022.08.031

Bernevega, A., & Gekker, A. (2021). The industry of landlords: Exploring the assetization of the triple-A game. *Games and Culture, 17*(1), 47–69. https://doi.org/10.1177/15554120211014151

Birch, K. (Ed.) (2020). *Assetization. Turning things into assets in technoscientific capitalism.* MIT Press.

Blackman, T., & Harley, D. (2024). Interpreting Apple's visions: Examining the spatiality of the Apple Vision Pro. *Platforms & Society, 1.* https://doi.org/10.1177/29768624241283913

Blackwood, G. (2023). Roblox and Meta Verch: A case study of Walmart's Roblox games. *M/C Journal, 26*(3). https://doi.org/10.5204/mcj.2958

Bleier, A., Fossen, B. L., & Shapira, M. (2024). On the role of social media platforms in the creator economy. *International Journal of Research in Marketing, 41*(3). https://doi.org/10.1016/j.ijresmar.2024.06.006

Bohné, T., Li, C., & Triantafyllidis, K. (2023). *Exploring the industrial Metaverse: A roadmap to the future.* World Economic Forum. Retrieved March 15, 2025, from https://www.weforum.org/publications/exploring-the-industrial-metaverse-a-roadmap-to-the-future/

Brachmann, E., & Prisacariu, V. A. (2024). *Building a large geospatial model to achieve spatial intelligence.* Retrieved March 15, 2025, from https://nianticlabs.com/news/largegeospatialmodel?hl=en

Browning, K. (2020, August 16). Where has your tween been during the pandemic? On this gaming site. *The New York Times.* https://www.nytimes.com/2020/08/16/technology/roblox-tweens-videogame-coronavirus.html

Cappa, M.L. (2025) *From Web 2.0 to the Metaverse: Analyzing the evolution of platform-based business models and the creator economy* [Master's thesis, Politecnico di Milano]. https://www.politesi.polimi.it/handle/10589/236207

Carter, R. (2024, December 27). What is project Aria and what does it mean for Meta's Orion glasses? *XR Today.* https://www.xrtoday.com/augmented-reality/what-is-project-aria-and-what-does-it-mean-for-metas-orion-glasses/

Carter, M., & Egliston, B. (2024). *Fantasies of virtual reality: Untangling fiction, fact, and threat.* MIT Press.

Challenor, J. (2023, October 12). Fortnite's new in-game Holocaust Museum shows us a virtual future for education. *The Conversation.* https://theconversation.com/fortnites-new-in-game-holocaust-museum-shows-us-a-virtual-future-for-education-215500

Cherney, M. (2021). Nvidia launches Omniverse for enterprise collaboration. *Protocol.* Retrieved March 15, 2025, from https://web.archive.org/web/20220518135750/https://www.protocol.com/enterprise/nvidia-omniverse-gtc-enterprise

Chess, S., & Consalvo, M. (2022). The future of media studies is game studies. *Critical Studies in Media Communication, 39*(3), 159–164. https://doi.org/10.1080/15295036.2022.2075025

Chia, A. (2022). The metaverse, but not the way you think: Game engines and automation beyond game development. *Critical Studies in Media Communication, 39*(3), 191–200. https://doi.org/10.1080/15295036.2022.2080850

Chia, A., Malazita, J. W., Young, C. J., Nieborg, D. B., Joseph, D. J., & Gantt, M. D. (2023). The engine is the message: Videogame infrastructure and the future of digital platforms. *AoIR Selected Papers of Internet Research, 2022.* https://doi.org/10.5210/spir.v2022i0.12954

Chow, A., & De Guzman, C. (2022, July 25). A crypto game promised to lift Filipinos out of poverty. Here's what happened Instead. *Time*. https://time.com/6199385/axie-infinity-crypto-game-phi lippines-debt/

Cieslak, M., & Gerken, T. (2023, December 7). Lego Fortnite: Gaming giant launches Minecraft rival. *BBC News*. https://www.bbc.com/news/technology-67635859

Couldry, N., & Meijas, U. A. (2019). *The costs of connection: How data is colonizing human life and appropriating it for capitalism*. Stanford University Press.

Cross, T. (2021, July 4). Roblox wants to build the Metaverse. Can it? *Wired*. https://www.wired.com/story/roblox-metaverse/

Cucco, J. (2022, February 20). Jump into the "Roblox" Metaverse for a new era of play. *The Toy Book*. https://toybook.com/jump-into-the-roblox-metaverse-for-a-new-era-of-play/

Delic, A. J., & Delfabbro, P. H. (2022). Profiling the potential risks and benefits of emerging "play to earn" games: A qualitative analysis of players' experiences with axie infinity. *International Journal of Mental Health and Addiction, 22*. https://doi.org/10.1007/s11469-022-00894-y

Duffy, B. E. (2017). *(Not) Getting paid to do what you love: Gender, social media, and aspirational work*. Yale University Press.

Egliston, B., & Carter, M. (2020). Oculus imaginaries: The promises and perils of Facebook's virtual reality. *New Media & Society, 24*(1), 70–89. https://doi.org/10.1177/1461444820960411

Egliston, B., & Carter, M. (2021a). Critical questions for Facebook's virtual reality: Data, power and the metaverse. *Internet Policy Review, 10*(4). https://policyreview.info/articles/analysis/cri tical-questions-facebooks-virtual-reality-data-power-and-metaverse

Egliston, B., & Carter, M. (2021b). Examining visions of surveillance in Oculus' data and privacy policies, 2014–2020. *Media International Australia, 188*(1), 52–66. https://doi.org/10.1177/132 9878X211041670

Egliston, B., & Carter, M. (2022). 'The interface of the future': Mixed reality, intimate data and imagined temporalities. *Big Data & Society, 9*(1), 1–15. https://doi.org/10.1177/205395172110 63689

Evans, L., Frith, J., & Saker, M. (2022). *From microverse to metaverse: Modelling the future through today's virtual worlds*. Emerald Group Publishing.

Foxman, M. (2019). United we stand: Platforms, tools and innovation with the Unity Game Engine. *Social Media + Society, 5*(4). https://doi.org/10.1177/2056305119880177

Foxman, M. (2021). Making the virtual a reality: Playful work and playbour in the diffusion of innovations. *Digital Culture & Society, 7*(1), 91–110. https://doi.org/10.14361/dcs-2021-0107

Foxman, M. (2022). Gaming the system: Playbour, production, promotion, and the metaverse. *Baltic Screen Media Review, 10*(2), 224–233. https://doi.org/10.2478/bsmr-2022-0017

Freedman, E. (2020). *The persistence of code in game engine culture*. Routledge.

Geyer, M. (2022, September 20). *On track: Digitale Schiene Deutschland building digital twin of rail network in NVIDIA Omniverse*. NVIDIA Blog. Retrieved March 15, 2025, from https://blogs.nvidia.com/blog/deutsche-bahn-railway-system-digital-twin/

Grimes, S. M. (2021). *Digital playgrounds. The hidden politics of children's online play spaces, virtual worlds, and connected games*. University of Toronto Press.

Google DeepMind. (2025, March 12). *Introducing Gemini Robotics and Gemini Robotics-ER, AI models designed for robots to understand, act and react to the physical world*. Retrieved March 15, 2025, from https://deepmind.google/discover/blog/gemini-robotics-brings-ai-into-the-phy sical-world/

Harley, D. (2020). Virtual bodies Inc. *Public, 30*, 250–259. https://doi.org/10.1386/public_00019_7

Hesselbein, C., Bory, P., & Canali, S. (2024). Six provocations for metaverse datafication: An emergent cultural, technological, and scholarly phenomenon. *Information, Communication & Society, 28*(5), 778–796. https://doi.org/10.1080/1369118X.2024.2433548

Hyman, J. (2021, November 12). Nvidia CEO: The metaverse will be "much, much bigger than the physical world." *Yahoo Finance*. https://finance.yahoo.com/news/nvidia-ceo-the-metaverse-will-be-much-much-bigger-than-the-physical-world-174256652.html

Irom, B. (2021). Virtual reality and celebrity humanitarianism: Rashida Jones in Lebanon. *Media, Culture & Society, 44*(1), 88–104. https://doi.org/10.1177/01634437211022725

Jebelli, H., Asadi, S., Mutis, I., Liu, R., & Cheng, J. (Eds.) (2024). *Digital twins in construction and the built environment*. American Society of Civil Engineers. https://doi.org/10.1061/978078448 5606

Joseph, D. (2021). Battle pass capitalism. *Journal of Consumer Culture, 21*(1), 68–83. https://doi.org/10.1177/1469540521993930

Jungherr, A., & Schlarb, D. B. (2022). The extended reach of game engine companies: How companies like Epic Games and Unity Technologies provide platforms for extended reality applications and the metaverse. *Social Media + Society, 8*(2), 1–12. https://doi.org/10.1177/205630512211 07641

Kastrenakes, J. (2017, October 9). A cartoon Mark Zuckerberg toured hurricane-struck Puerto Rico in virtual reality. *The Verge*. https://www.theverge.com/2017/10/9/16450346/zuckerberg-facebook-spaces-puerto-rico-virtual-reality-hurricane

Kerr, A. (2017). *Global games: Production, circulation and policy in the networked era*. Routledge.

Kreuzer, T., Papapetrou, P., & Zdravkovic, J. (2024). Artificial intelligence in digital twins—A systematic literature review. *Data & Knowledge Engineering, 151*, 102304–102304. https://doi.org/10.1016/j.datak.2024.102304

Kücklich, J. (2005). Precarious playbour: Modders and the digital games industry. *The Fibreculture Journal, 5*(1). https://five.fibreculturejournal.org/fcj-025-precarious-playbour-modders-and-the-digital-games-industry/

Li, F.-F. (2024, November 20). Fei-Fei Li says understanding how the world works is the next step for AI. *The Economist*. https://www.economist.com/the-world-ahead/2024/11/20/fei-fei-li-says-understanding-how-the-world-works-is-the-next-step-for-ai

Lv, Z. (2023). Generative artificial intelligence in the metaverse era. *Cognitive Robotics, 3*(1), 208–217. https://doi.org/10.1016/j.cogr.2023.06.001

Magaudda, P. (2015). Apple's iconicity: Digital society, consumer culture and the iconic power of technology. *Sociologica, 9*(1). https://doi.org/10.2383/80397

McArthur, V. (2024, September 23). Meta's AI-powered smart glasses raise concerns about privacy and user data. *The Conversation*. https://theconversation.com/metas-ai-powered-smart-glasses-raise-concerns-about-privacy-and-user-data-238191

Messeri, L. (2024). *In the land of the unreal: Virtual and other realities in Los Angeles*. Duke University Press.

Meta (2021) *Founder's letter, 2021*. Retrieved March 15, 2025, from https://about.fb.com/news/2021/10/founders-letter/

Meta. (2025, February 27). *Introducing Aria Gen 2: Unlocking new research in machine perception, contextual AI, robotics, and more*. Retrieved March 15, 2025, from https://www.meta.com/en-gb/blog/project-aria-gen-2-next-generation-egocentric-research-glasses-reality-labs-ai-robotics/

Milk, C. (2015). *How virtual reality can create the ultimate empathy machine*. TED Talks (video). https://www.ted.com/talks/chris_milk_how_virtual_reality_can_create_the_ultimate_empathy_machine?language=en

Morozov, E. (2022, January 13). Web3: A map in search of territory. *The Crypto Syllabus*. https://the-crypto-syllabus.com/web3-a-map-in-search-of-territory/

Nair, V., Munilla Garrido, G., Song, D., & O'Brien, J. F. (2023). Exploring the privacy risks of adversarial VR game design. *Proceedings on Privacy Enhancing Technologies, 2023*(4), 238–256. https://doi.org/10.56553/popets-2023-0108

Nakamura, L. (2020). Feeling good about feeling bad: Virtuous virtual reality and the automation of racial empathy. *Journal of Visual Culture, 19*(1), 47–64. https://doi.org/10.1177/147041292 0906259

Nieborg, D. B., & Poell, T. (2018). The platformization of cultural production: Theorizing the contingent cultural commodity. *New Media & Society, 20*(11), 4275–4292. https://doi.org/10.1177/1461444818769694

Nvidia (n.d.) A timeline of innovation. About us. Retrieved March 15, 2025, from https://www.nvi
    dia.com/en-eu/about-nvidia/corporate-timeline/
Ongweso Jr, E. (2022, April 4). The metaverse has bosses too. Meet the "managers" of
    Axie Infinity. *VICE*. https://www.vice.com/en/article/the-metaverse-has-bosses-too-meet-the-
    managers-of-axie-infinity/
Parkin, S. (2022, January 9). The trouble with Roblox, the video game empire built on child labour.
    *The Guardian*. https://www.theguardian.com/games/2022/jan/09/the-trouble-with-roblox-the-
    video-game-empire-built-on-child-labour
Reality Labs. (2019, September 25). *Inside Facebook reality labs: Research updates and the future
    of social connection*. Facebook. Retrieved March 15, 2025, from https://tech.facebook.com/
    reality-labs/2019/9/inside-facebook-reality-labs-research-updates-and-the-future-of-social-con
    nection/
Roblox (n.d.). *Experiences and places*. Retrieved March 15, 2025, from https://create.roblox.com/
    docs/production/publishing/publish-experiences-and-places
Sadowski, J. (2019). When data is capital: Datafication, accumulation, and extraction. *Big Data &
    Society, 6*(1), 1–12. https://doi.org/10.1177/2053951718820549
Sadowski, J. (2020). The internet of landlords: Digital platforms and new mechanisms of rentier
    capitalism. *Antipode, 52*(2), 562–580. https://doi.org/10.1111/anti.12595
Saker, M., & Frith, J. (2020). Coextensive space: Virtual reality and the developing relationship
    between the body, the digital and physical space. *Media, Culture & Society, 42*(7–8), 1427–1442.
    https://doi.org/10.1177/0163443720932498
Sato, M. (2022, September 9). Roblox is ready to grow up. *The Verge*. https://www.theverge.com/
    2022/9/9/23343459/roblox-age-guidelines-metaverse-ads-developer-conference-announcem
    ents?scrolla=5eb6d68b7fedc32c19ef33b4
Seif El-Nasr, M., Drachen, A., & Canossa, A. (2013). *Game analytics: Maximizing the value of
    player data*. Springer.
Shah, S. F. A., & Sukthankar, G. (2015). Mining social interaction data in virtual worlds. In F.
    Koch, F. Meneguzzi, & K. Lakkaraju (Eds.), *Agent technology for intelligent mobile services
    and smart societies* (pp. 86–105). Springer. https://doi.org/10.1007/978-3-662-46241-6_8
Shapiro, E. (2021, April 18). Nvidia CEO Jensen Huang talks the powers of automation. *Time*.
    https://time.com/5955412/artificial-intelligence-nvidia-jensen-huang/
Smith, H. (2024). The metaverse-industrial complex. *Information, Communication & Society, 28*(5),
    797–814. https://doi.org/10.1080/1369118X.2024.2423346
Solon, O. (2017, October 10). Mark Zuckerberg "tours" flooded Puerto Rico in bizarre virtual
    reality promo. *The Guardian*. https://www.theguardian.com/technology/2017/oct/09/mark-zuc
    kerberg-facebook-puerto-rico-virtual-reality
Stockdale, H. (2024, July 2). Tim Sweeney explains why Fortnite has "no plans" for
    VR support. *UploadVR*. https://www.uploadvr.com/tim-sweeney-explains-why-fortnite-has-
    no-plans-for-vr-support/
Swisher, K. (2021, November 11). Opinion: The Metaverse: Expectations vs. reality. *The New York
    Times*. https://www.nytimes.com/2021/11/11/opinion/sway-kara-swisher-jaron-lanier.html
Takahashi, D. (2024, October 28). Epic Games CEO Tim Sweeney's path to the open metaverse is
    via enlightened self-interest. *VentureBeat*. https://venturebeat.com/ai/epic-games-ceo-tim-swe
    eneys-path-to-the-open-metaverse-is-via-enlightened-self-interest/
Thatcher, J., O'sullivan, D., & Mahmoudi, D. (2016). Data colonialism through accumulation by
    dispossession: New metaphors for daily data. *Environment and Planning d: Society and Space,
    34*(6), 990–1006. https://doi.org/10.1177/0263775816633195
The Economist. (2019, November 21). Venezuela's paper currency is worthless, so its people seek
    virtual gold. *The Economist*. https://www.economist.com/the-americas/2019/11/21/venezuelas-
    paper-currency-is-worthless-so-its-people-seek-virtual-gold
The Economist. (2023, March 20). Ready, player four billion: the rise of video games. *The
    Economist—Special Report Video Games*. https://www.economist.com/special-report/2023/03/
    20/ready-player-four-billion-the-rise-of-video-games

Tulleken, H. (2019). Four decades of advergames. *Gamasutra* [Now *Game Developer*]. Retrieved March 15, 2025, from https://web.archive.org/web/20190613165924/https://www.gamasutra.com/blogs/HermanTulleken/20190612/344674/Four_Decades_of_Advergames.php

Veix, J. (2019) Exploring the digital ruins of Second Life. *digg*. Retrieved March 15, 2025, from https://web.archive.org/web/20250228065915/https://digg.com/2018/second-life-in-2018

Wakefield, J. (2021, March 10). Roblox: How the children's game became a $30bn bet on the Metaverse. *BBC News*. https://www.bbc.com/news/technology-56345586

Webster, A. (2020, July 1). The latest modern art installation is inside Fortnite. *The Verge*. https://www.theverge.com/2020/7/1/21308391/fortnite-creative-manchester-international-festival-art-exhibition-laturbo-avedon

Webster, A. (2022, April 11). Epic announces $2 billion in funding for its metaverse efforts. *The Verge*. https://www.theverge.com/2022/4/11/23020134/epic-2-billion-funding-metaverse-sony-lego

Webster, A. (2024, February 8). Fortnite is winning the metaverse. *The Verge*. https://www.theverge.com/24065901/fortnite-metaverse-disney-epic-partnership

Wiener, A. (2022, January 4). Money in the Metaverse. *The New Yorker*. https://www.newyorker.com/news/letter-from-silicon-valley/money-in-the-metaverse

Wiener, A. (2024, April 15). How perfectly can reality be simulated? *The New Yorker*. https://www.newyorker.com/magazine/2024/04/22/can-the-world-be-simulated

Xu, T. (2022, November 24). We could run out of data to train AI language programs. *MIT Technology Review*. https://www.technologyreview.com/2022/11/24/1063684/we-could-run-out-of-data-to-train-ai-language-programs/

Yeo, S. (2024, November 21). Tuvalu: The disappearing island nation recreating itself in the metaverse. *BBC News*. https://www.bbc.com/future/article/20241121-tuvalu-the-pacific-islands-creating-a-digital-nation-in-the-metaverse-due-to-climate-change

Zuboff, S. (2019). *The age of surveillance capitalism: The fight for a human future at the new frontier of power*. Public Affairs.

# Chapter 4
# Metaverse Bodies

**Abstract** This chapter examines the ways in which metaversal technologies, such as VR and AR devices as well as virtual worlds and gaming platforms, seek not only to collect new and more extensive forms of data about bodies, but also to extend as well as narrowly redefine our corporeal capacities and individual identities. In particular, we focus on how metaverse technologies and worlds are positioned as new spaces for the expression of identity through the consumption of virtual fashion and apparel while simultaneously commodifying existing forms of social interaction, economic production, and self-presentation. Last, we reflect on some of the connections and implications of the virtualization and datafication of bodies, particularly for the development of AI systems that are becoming increasingly present in virtual as well as physical environments.

## 4.1 Being Inside the Metaverse

As noted at the start of the previous chapter, dominant definitions of the metaverse portray this as an immersive, 3D environment in which people, represented by virtual avatars, interact with each other as well as carry out a wide variety of professional and leisure activities. One of the main actors shaping this definition is Meta, the company that not only runs the Horizon Worlds VR platform but also produces popular mixed-reality headsets and, together with EssilorLuxottica, smart glasses that are also starting to become a commercial success (Rodriguez & Vanian, 2024). When talking about the metaverse, Mark Zuckerberg frequently stresses how people, social interactions, and presence are its defining aspects. For example, in an interview given in early 2021 before the infamous name change from Facebook to Meta, Zuckerberg stated that

> "one of the things that I really wanted to build was basically the sense of an embodied internet where you could be in the environment and teleport to different places and be with friends [...] in a way that's more natural and lets us feel more present with people" (Newton, 2021).

During the announcement of the company's pivot late that year, the metaverse was presented as "an embodied internet where you're in the experience" (Meta, 2021a).

© The Author(s) 2025
C. Hesselbein and P. Bory, *Infrastructures of Reality: Metaverse Stories, Spaces, Bodies*,
PoliMI SpringerBriefs, https://doi.org/10.1007/978-3-031-97167-9_4

Zuckerberg is certainly not the only one to emphasize the importance of sociality and corporeal practices in 3D virtual environments. Jensen Huang of Nvidia has claimed that "we'll be able to almost feel like we're there with each other" (Takahashi, 2021), and Tim Sweeney of Epic has similarly asserted that "you and your friends and your appearance and cosmetics can go from place to place and have different experiences while remaining connected to each other socially" (Herrman & Browning, 2021).

Embodiment, identity, and social interaction are thus thought to be core aspects of the metaverse, another example of the ongoing attempt by large technology companies to collapse the boundaries between physical and virtual realms and to encroach on and subsume activities that are generally done 'offline'. As such, the metaverse can be understood as a technological means to "extend embodied social presence" and interpersonal communication across "spatial–temporal barriers" in a manner that resembles conventional face-to-face interaction (Evans et al., 2022). It is precisely for this reason that companies are making such great efforts to introduce new forms of virtual identity and social interaction that mimic the ease, comfort, and pleasure of communicating and performing social identity in the 'real' world. However, such new modes of sociality also come with the threat of reproducing or exacerbating existing negative forms of social interaction, such as harassment and discrimination, and moreover, are open to commodification and exploitation by the companies that create and operate such virtual environments and devices.

This chapter continues our exploration of how the metaverse can be understood as an attempt to virtualize and datafy the world, but we now turn from spaces to bodies. To examine how the metaverse and the companies driving this endeavour are involved in the virtualization and datafication of bodies and corporeal practices, Meta is, once again, a good starting point not only because of its leading role in the development of commercial AR and VR devices, but especially because of its history as a company that has exploited and financially benefited from the data of the users on its various platforms. And, more than perhaps any other company, Meta appears to have gone all in on betting its future on the creation of the 'Metaverse'. The company's model for conceptualizing and drawing connections between embodiment and the metaverse is thus likely to be a highly influential blueprint for other companies to follow, and will likely shape how metaversal technologies and their relationship with space and embodiment will be co-constructed. The concept of space—both to describe platforms or 'microverses' as well as the varied places or games on platforms—will remain important for this chapter because any discussion of spaces and bodies needs to consider both simultaneously (Osborne & Jones, 2022). Our emphasis here, however, will be on how embodiment and identity are conceptualized in relation to the metaverse, how embodied practices and the performance of identity are being drawn into and shaped by virtual spaces, and how such practices and bodies are being virtualized and datafied through metaversal technologies.

## 4.2 The Paradox of Meta's Metaverse

During its annual Facebook Connect conference on October 28, 2021, Facebook announced the change of its name to Meta and laid down its vision of the metaverse (Meta 2021a; 2021b). During the almost 80-min-long keynote speech—named "The Metaverse and How We'll Build It Together"—Mark Zuckerberg together with several senior colleagues expounded to audiences in the room as well as multiple online venues what the metaverse is and will be (Meta, 2021c). As the next "platform and medium", the metaverse and the various technologies Meta is working on will, it is claimed, enable deeper social connections, provide a greater variety of entertainment (particularly through gaming), enable more productive work environments, especially as "creators and developers" in a new virtual economy, but also offer assistance with physical exercise as well as facilitate new and more personalized forms of education and professional training. At the centre of this interconnected world of virtual apps and environments that appear seamlessly enmeshed with both the physical world and human bodies is the avatar, the "living 3D representations of you, your expressions, your gestures that are going to make interactions much richer than anything that's possible online today". In short, Meta's metaverse will lead us—in the next five or ten years—into a realm of near infinite possibilities, a future "beyond anything we can imagine".

Unlike the 'visit' to hurricane-struck Puerto Rico described in the previous chapter, Zuckerberg's presentation of Meta's future as a metaverse company did not generate public controversy, although it did attract some derision for its awkward presentation style, zany visuals, grandiose vision, and techno-utopian tenor. Instead, it set off an intense if relatively short-lived hype about metaverses as well as ancillary phenomena and technologies, such as NFTs, blockchains, and cryptocurrencies. Meta's turn to the metaverse is therefore widely regarded as a pivotal moment in the development of virtual worlds, not least because of the enormous amount of technical and financial resources that a company the size of Meta can afford to dedicate to developing the metaverse. Meta's version of the 'Metaverse', moreover, is particularly ambitious and all-encompassing, not just in a technical sense as an interconnected and interoperable 'world' but especially in a socio-cultural and economic sense. Meta's metaverse seeks to position itself as an infrastructure in which social interactions, entertainment, education, exercise, commerce, and work can all be engaged with or carried out. In short, pretty much anything an average human would do throughout their day, but now 'better' because these activities are somehow more engaging, fulfilling, and pleasurable once carried out in the metaverse.

This totalizing if not utopian vision of the metaverse and its ostensibly transformative significance appears to have been circulating within the company for some time. For instance, just after the 2021 Connect conference, reporting emerged about a 2018 pitch sent by Jason Rubin—then still an Oculus VR executive, now the Vice-President of Metaverse Experience at Meta—to Facebook executive board member and leading venture capitalist, Marc Andreessen, laying out a similar vision of the metaverse. In his pitch, Rubin describes future occupants of the metaverse as follows:

"the only thing she spends as much time doing as she spends in the Metaverse is working, eating, socializing, and sleeping in the IRL [in real life] 'MEATverse'". Note not only how the metaverse is situated as a place where people will spend almost all of their daily time, but also how this disembodied virtual space is juxtaposed with the fleshiness of 'real life'. What is more, Rubin suggests that "I might check in to Facebook multiple times a day, but I will LIVE in the Metaverse, work in the Metaverse, and potentially prefer my time in the Metaverse to my day-to-day grind" (Rodriguez, 2021). In other words, the metaverse is as much an escape from the daily 'grind' as it is a new, apparently disembodied, and ostensibly better place to inhabit and live.

Rubin's representation of the metaverse as an escape from mundane, embodied reality harkens back to previous ideas articulated in the 1990s about 'cyberspace' as a disembodied realm in which one can behave and live more freely (Hayles, 1999). Indeed, John Perry Barlow's famous declaration of independence (see Chap. 2), claims that "[cyberspace] is a world that is both everywhere and nowhere, but it is not where bodies live", and that here our identities "have no bodies" (1996). Moreover, fantasies of escaping embodiment and indeed mortality appear to be relatively widespread among the communities of engineers and entrepreneurs in Silicon Valley (Gebru & Torres, 2024; Messeri, 2024), and researchers have observed that metaverse narratives frequently contain deep-seated assumptions about virtuality as separated from corporeality (Kalpokas & Kalpokienė, 2024). It is perhaps no surprise, then, that some of these ideas have rubbed off on Andreessen, a successful 'tech' entrepreneur who has played a paramount role in the digital media landscape since the birth of the World Wide Web in the 1990s, which is when he co-launched the Mosaic browser, an early so-called killer app of the Web (Barcella et al., 2025). Notably, Andreessen is also an early investor in Facebook and OpenAI as well as a slew of cryptocurrency companies. Indeed, he is a particularly important actor in forwarding techno-utopian discourse—see for instance his 'techno-optimist manifesto' (Andreessen, 2023)—as well as in shaping technological development as the co-founder of Andreessen Horowitz, an influential Silicon Valley venture capital firm. Asked in an interview in early 2021 about whether digital technologies might make us 'too connected', Andreessen responds, rather tellingly and therefore worth quoting at length, as follows:

> A small percent of people live in a real-world environment that is rich, even overflowing, with glorious substance, beautiful settings, plentiful stimulation, and many fascinating people to talk to, and to work with, and to date. These are also *all* of the people who get to ask probing questions like yours. Everyone else, the vast majority of humanity, lacks Reality Privilege -- their online world is, or will be, immeasurably richer and more fulfilling than most of the physical and social environment around them in the quote-unquote real world. The Reality Privileged, of course, call this conclusion dystopian, and demand that we prioritize improvements in reality over improvements in virtuality. To which I say: reality has had 5,000 years to get good, and is clearly still woefully lacking for most people; I don't think we should wait another 5,000 years to see if it eventually closes the gap. We should build -- and we are building -- online worlds that make life and work and love wonderful for everyone, no matter what level of reality deprivation they find themselves in. (Soldo, 2021)

Here, the 'real' or 'material' world—Rubin's 'meatspace'—is portrayed by Andreessen as fundamentally unequal and unfulfilling, and virtual worlds are positioned as solutions to such social issues because they will apparently make life and work 'immeasurably richer and more fulfilling'. Virtual technologies and spaces allow us, in other words, to escape from the natural or 'real' state of inequality towards a virtual world of abundance. That is, the metaverse is posited as a virtual silver bullet for solving the widespread problems of 'reality', whether these may be the shortcomings of embodiment or the deprivations of society. In short, the realm of virtual worlds is posited as the ultimate 'technological solutionist' answer to social inequality (Morozov, 2013).

Despite these circulating claims about the supposed disembodied nature of the metaverse, Meta's presentation of the metaverse at the Connect 2021 conference nevertheless contains multiple explicit references to embodied practices. Zuckerberg frequently employs corporeal terms and metaphors, such as when he refers to 'jumping into apps' or cracks jokes about needing 'more sunscreen', and spends considerable time explaining how Quest headsets and virtual spaces will allow people to exercise and 'work out in new worlds, even against an AI'. In Zuckerberg's background, sports equipment such as bicycles and surfboards frequently appear, and he is portrayed as enjoying surfing and fencing in the metaverse. Furthermore, when discussing the commercial potential of the metaverse, great emphasis is placed on embodied aspects of avatars and how their appearance can be modified with digital clothing. And, in a seemingly unexpected turn, Jackie Aina, a popular beauty influencer who is best known for creating makeup, perfumes, and candles—i.e. physical products involving haptic and olfactory sensations that are not (yet) possible to convey through metaverse technologies—is introduced to demonstrate how the metaverse can be a commercial opportunity for existing businesses. In short, Zuckerberg's rendition of the metaverse appears quite full of embodied experiences and sensory pleasures. Consequently, a contradiction emerges between representations of the metaverse as either an embodied reality or an escape from this (Saker & Frith, 2022). The metaverse thus appears in a paradoxical relationship with embodiment because it is as much a place to virtually depart from one's corporeality as it is a place to engage in deeply corporeal activities.

## 4.3 From Embodiment to Experience, or from Meatspace to Metaspace

This seeming contradiction, however, is perhaps less paradoxical if one notes that, in addition to references to embodiment, Zuckerberg makes even more references to affective and experiential states. In the first five minutes of the presentation, for example, Zuckerberg employs the word 'feel' numerous times, describing how the metaverse will make you "feel…right in the moment" or "feel like you're right there". Throughout the presentation, 'feel' or 'feelings' are brought in relation with music, people, facial expressions, and even more broadly, with the sense of seamless connection that the metaverse is supposed to offer. Indeed, "the feeling of presence" is emphasized as "the defining quality of the metaverse". Even more telling is the ubiquitous use of the term 'experience', which, as already noted in the previous chapter, is a term that companies employ both for describing metaversal spaces as well as what happens in such spaces. During the Connect 2021 presentation, Zuckerberg uses 'experience' over 60 times to denote at least three distinct yet also overlapping phenomena that range from clearly material and embodied to ostensibly immaterial and disembodied. These different uses of 'experience' are also found in the documentation released by Meta around the period of the company's name change (Meta, ), in descriptions on the Meta Store, as well as in various press releases, interviews, and so on. First, 'experience' is employed in the commonsense meaning of a state in which a person is being affected through perception of or participation in an event or moment. For instance, people 'experience the world and interact with each other'. Second, 'experience' is used to denote various media that capture 'experiences' in the first sense and that can subsequently be shared with others as, for instance, a picture of video. Here 'experience' is conceptualized as a type of content that can be circulated on social media platforms or through apps. Third and most importantly, 'experience' is used to denote the state of using or being 'inside' an app, technology, platform, or virtual world. For example, the "metaverse is an embodied internet where you're in the experience" or "you should be able to bring your avatar and digital items across different apps and experiences in the metaverse". In other words, the term 'experience' is employed to cover a range of different perceptual moments, emotional states, embodied practices that are enabled by technological objects, infrastructures, and environments. What is more, these three similar yet importantly distinct meanings of the term are used in a manner that suggests corporeality—all experiences, after all, arise from the body and its interactions with the surroundings—while nonetheless retaining a crucial and somewhat ambiguous emphasis on intangible or immaterial aspects.

The language of 'experience' as employed by Meta, as well as other metaverse companies, such as Apple and Roblox, thus subtly elides the boundaries between being embodied or disembodied, between feeling and being, as well as between engaging in material or immaterial practices. Talk of feelings and experiences allows Zuckerberg to refer to people without explicitly addressing that they have bodies or that feelings or experiences are always necessarily embodied. Although Meta claims to deliver 'immersive experiences', these are always embodied and therefore susceptible to datafication (Lupinacci, 2022). Bodies, affects, and technologies are, in short, seamlessly brought together in this specific metaverse narrative. And above all, this narrative obfuscates that such embodied practices and experiences are mediated through Meta's headsets or glasses, taking place in Meta's applications and on their platforms, and that these behaviours and experiences are therefore open to being datafied, commodified, and exploited in manners that are likely to go far beyond the company's current and already infamous reputation.

The conceptual distinctions between embodiment and disembodiment, materiality and immateriality, as well as being 'online' or 'offline' have long been a central point of contention in discussions about digital technologies and the internet (Baym, 2010; Hayles, 1999; Turkle, 1995). The metaverse, as an ostensible attempt to bridge, once and for all, the gap between these binary categories, presents perhaps the apex of this complicated entanglement. Or put differently, the metaverse can be understood as an instance in which such binaries can no longer be upheld. We are, after all, already always online and offline, our bodies and identities distributed across physical and digital spaces, and our cultural forms are both material and virtual. Although these terms are distinct—and can be used to analytically separate certain aspects of various phenomena—they are also inherently connected. Put more pithily, "both the virtual and the physical are real" (Boellstorff, 2015, xxi). Moreover, making connections between physical and virtual practices, and therefore different types of data, is something that technology companies are deeply invested in because it allows for their aggregation and analysis in a manner than can provide deeper insights. Google's purchase of wearable manufacture Fitbit is one prominent example, but Meta's initial attempts at linking Facebook profiles (already governed by the companies 'real name' policy) with Quest accounts, let alone how the data flowing from such spaces can be brought together, is also a matter of concern in terms of surveillance and datafication as our so-called 'contiguous identities' provide a connection between virtual and physical realms (Saker & Frith, 2022).

In the context of VR technologies, often interpreted as a primary metaverse technology, these problematic distinctions rear their head once more, particularly in the discourse around VR as the ultimate empathy machine, which is posited as engendering feelings and experiences that are otherwise difficult or impossible to access. As introduced in the previous chapter, VR is conceptualized as having the ability to connect people in a more empathetic manner than conventional media can achieve, and therefore of having a specific potential for doing 'good'. The idea that new forms of media can allow humans to better imagine or empathize with the emotional world of other humans is not new (Heft-Luthy, 2019), but in the case of VR, has widely and enthusiastically been taken up by journalists, filmmakers, and technology companies. For instance Oculus, when still a subsidiary VR company of Facebook, funded the VR for Good initiative, which states that it "fosters and promotes immersive storytelling, focused on social impact […] to create human-centric stories that promote empathy and empowerment" (Meta, n.d.), and that has subsequently released a slew of VR films that focus on issues related to racial and social justice. Needless to say, the idea that VR can be a powerful technology not just for more engaged and interactive forms of storytelling across various media but also for achieving 'good' has provided an important argument for companies such as Meta to further justify the development of VR technology and to assert its relevance to practices beyond gaming. The 'VR for Good' discourse is also another potent, and problematic, example of technological solutionism that feeds into escapist fantasies surrounding the metaverse discussed earlier in this chapter.

A key notion in the conceptualization of VR technology as an empathy machine is, once again, that of 'experience', and particularly the feeling or experience of 'presence'. On the one hand, the experience of presence is predicated on the idea that VR can allow people to be virtually present in places that they physically are

not. Indeed, as Zuckerberg states in the letter' announcing Facebook's name change
to Meta:

> The defining quality of the metaverse will be a feeling of presence — like you are right
> there with another person or in another place. Feeling truly present with another person is
> the ultimate dream of social technology. (Meta, 2021a)

As Zuckerberg's 'trip' to Puerto Rico described in the previous chapter indicates,
such ideas about virtual co-presence, and a supposed associated increase in levels
of empathy, are not convincing. On the other hand, the notions of 'experience' and
'empathy' are also predicated on the idea that VR can allow one to 'inhabit another
body' and thus allow one "to see through another's eyes, embodying their experiences, thus 'empathising' with them" (Bollmer, 2017, 63). VR video technologies,
such as 360° cameras, are thought to facilitate "special affects" (a twist on the cinematic technique of 'special effects') because they create an illusion for viewers of
being present in the scene together with the other characters in that scene. It is this
"experience of 'being with' that lends credence to the fantasy of 'being someone
else'" (Messeri, 2024, 147). By temporarily seeming to inhabit another person's
world, and therefore briefly embodying their perspective and living their experiences,
such 'special affects' are claimed to translate into an improved understanding of and
empathy for other people's lives. VR thus appears to facilitate embodied and affective access to the realities experienced by others, and to provide such an experiential
equivalent in an apparently 'frameless' or 'unmediated' manner (Carter & Egliston,
2024). Indeed, as Zuckerberg puts it in his keynote speech, "instead of looking at
a screen, you're going to be in these experiences" (Meta, 2021c), thus precisely
suggesting that the distance between screen and viewer has entirely collapsed. Note,
once again, the subtle shift from being an embodied actor looking at a material object
to having a virtual 'experience' that is simultaneously a place to be in, or a form of
media, as well as an experiential state. In short, a crucial ambivalence is maintained
between having a physical body and embodying a virtual avatar.

   The conceptualization of the body and embodiment in the metaverse as presented
above by Zuckerberg and Andreessen as well as those who buy into the 'VR for
good' discourse, as inspiring as it may be at times, runs into several problems. First,
inhabiting another's body does not mean inhabiting their stories, their memories,
and their past experiences. No matter how complete an immersive experience may
be from a sensory perspective, identification and empathy with any 'other' requires
not only a body and a space, but also a shared time—if not an entire life experienced
in the same way. Second, one cannot actually leave one's body 'at the door' of
a virtual space (Penny, 1993). Rubin and Andreessen's overflowing and plentiful
virtual world, regardless of whether this can actually be accomplished through the
creation of metaversal spaces, will inevitably remain tethered to 'meatspace' and its
fleshy bodies. Moreover, bodies and minds as well as embodiment and affect are
inextricably connected. And last but certainly not least, metaversal 'experiences'
will remain firmly rooted in the very material experiences of the bodies wearing
Meta's headsets or glasses, not to mention their subjection to the company's business
operations, computational infrastructures, and datafication processes.

However, the birth of Meta as a company and Zuckerberg's vision of the metaverse should not be seen as the only version of the metaverse, particularly because Meta is not the only company developing virtual worlds. Furthermore, Meta's conceptualization of the metaverse hinges strongly on the use of VR even if the company now seems to be moving towards hybrid definitions that emphasize mixed reality rather than only VR. More importantly, other platforms, such as Second Life, Minecraft, Roblox, and Fortnite have been around for considerably longer and attract many more people compared to Meta's Horizon Worlds. And gaming platforms in particular have established themselves as new arenas for the performance of identity. As noted in the previous chapter, such virtual spaces are becoming increasingly successful at drawing in diverse activities and practices that were previously carried out elsewhere, such as new forms of labour and ownership. Here, we will examine a variety of currently existing gaming platforms or 'microverses' and discuss how these seek to redefine, reshape, and commodify and datafy virtual embodiment and identity according to their specific visions and interests.

## 4.4 Gaming Platforms and the Fashioning of Virtual Identity

One prime example of an activity that is currently being drawn in or extended into virtual worlds is the performance of social identity. That identities have started to become digital or virtual is, to be sure, not an entirely new process. Digital governance initiatives and social media platforms are but two obvious examples of how our identity has become digitally mediated over the past two decades if not longer. However, Evans and colleagues also point out that the metaverse can be defined from the perspective of people who engage in virtual practices rather than from the perspective of virtual platforms or spaces, namely as "when your virtual self becomes more important than your physical self" (Evans et al., 2022, 4). To what extent this can truly be said to be the case is, of course, debatable. On the one hand, some people already value their 'online' identities more than their 'offline' ones, regardless of the existence of virtual worlds, and in some ways we already spend more time interacting with more people through digital or virtual channels than through face-to-face interaction. On the other hand, it is unclear whether those who are very active on, say, social media or gaming platforms and in virtual worlds really value their 'digital' or 'virtual' identities more than their physical ones.

The truth is likely to lie somewhere in the middle. After all, we all already have multiple identities that we value and perform to lesser or greater extents depending on the context, regardless of whether this is 'online' or 'offline' (Baym, 2010; Boyd, 2014; Papacharissi, 2011). However, what does seem to be clear is that both 'Big Tech' companies and gaming companies as well as larger groups of people, often referred to as users or gamers, are placing greater emphasis on and investing more resources in the performance of digital or virtual identities. And platform companies in particular are seeking to position themselves as places to enact and especially invest in the expression of one's identity. As noted at the beginning of this chapter, Meta heavily emphasizes that the metaverse is a place for social interactions conducted via avatars, and Epic similarly stresses how social interactions as well as one's 'appearance and cosmetics' will be at the centre of virtual worlds (Herrman & Browning, 2021). And many people indeed appear to be investing considerable amounts of time and money in fashioning their virtual identities. In other words, much like the development of virtual spaces allows for the creation of new forms of labour and datafication, the creation of virtual identities enables the creation of new forms of consumptions and commodification.

At the centre of the performance of identity in virtual worlds is the figure of the avatar. This figure is importantly different to the way identity is enacted on social media platforms, which can be understood as a previous iteration by 'Big Tech' companies to position themselves as a means for expressing identity. Rather than the profile pictures, text descriptions, and reams of photos and videos that we have come to expect on social media platforms, in virtual worlds you are represented by an animated, moving character that one uses or 'embodies' to navigate in a game (e.g. Fortnite), or between and in multiple games (e.g. Roblox), or to represent oneself while interacting with others in a virtual environment (e.g. VRChat, Microsoft Mesh). As one can tell from these examples, it is not strictly necessary to employ VR technologies in order to employ or embody an avatar. In VR, however, the avatar's movements are intended to correspond to the movements of the body of the person wearing the VR headset, which means that the relationship between one's body and avatar are experienced in a more direct and intimate manner. That being said, identifying oneself with an avatar or game character can happen just as easily without the employment of VR technologies. Moreover, as Boellstorff points out (2015), 'virtual reality' and 'virtual worlds' are not the same thing. Whereas the former signifies an interface involving sensory immersion that can potentially be used to access a virtual world, the latter involves social immersion among other people and their lives. In other words, 3D or even photorealism are not requirements at all in order for people to participate in social activities on, say, Roblox or Minecraft.

As noted in the previous chapter, and as witnessed from Meta's Connect 2021 presentation, one of the supposed promises of life in the metaverse is near-infinite customizability. One's metaverse home, according to Zuckerberg, will allow you to explore opportunities beyond one's imagination, such as having an "incredibly inspiring view" and "things that are only possible virtually" (Meta, 2021c). Similarly, virtual work environments will offer "room customization" and provide "your perfect work setup and you can actually do more than you could in your regular work setup". And in a rare reference to actual, physical embodiment, Zuckerberg asserts that "on top of all that you can keep wearing your favourite sweatpants". In a similar vein, gaming platforms such as Roblox and Minecraft, and to some extent Fortnite, allow one to construct virtual worlds with seemingly endless options. Similar to 'cyberspace', one of the dominant logics of virtualization is, after all, that this allows one to move beyond the apparent restrictions and limitations of physical reality and to enjoy personalized, tailored experiences in virtual worlds. And the ostensibly freer characteristics of virtual spaces are posited as a powerful alternative for carrying out one's daily activities compared to the limitations set by 'real-world' places or 'meatspace'.

The logics of near-limitless personalization and customization as well as the purported escape from physical reality are, of course, also extended to the body and one's social identity. Although the body has to be 'left at the door' in virtual spaces (Penny, 1993), one's identity does not. That is to say, the emergence of digital platforms and virtual technologies, over the past three decades, has given rise to discussions about how the performance of identity is shifting in online spaces, and particularly through social media platforms and other spaces where one's self-identity can be considered relevant. Company discourse about avatars presents this virtual representation of the self as an opportunity for playful experimentation and creative self-expression, above all through digital fashion and accessories (e.g. Meta, 2022). Left generally unmentioned here is that virtual identities also create a great commercial opportunity for commodification by the companies that operate metaverse environments. Practices of consumption, particularly in the form of 'lifestyle choices' are, after all, one of the primary means of expressing one's identity in contemporary capitalist contexts (Miller, 2012). Virtual worlds thus open up a new frontier for both the expression of identity as well as the commodification of this process, particularly through the production and consumption of virtual items, accessories, apparel, and emotes (Costa Pinto et al., 2015). And, of course, these behaviours and practices as well as 'engagement' with such virtual forms of expression can provide yet another valuable source of data.

Clothing and its communicative and expressive properties are a crucial aspect of both individual and collective culture, particular in terms of one's embodiment and self-identity, and it is therefore no surprise that this cultural role is valued, reproduced, and indeed extended and elaborated upon in virtual worlds. Indeed, the convergence of the fashion and gaming industries around and in virtual worlds as well as the recent emergence of virtual economies that revolve around the customization and beautification of the avatar is one important indication of how these worlds have managed to position themselves as a meaningful social context for the enactment of identity. For example, although physical clothing has long been referred to as the 'second' or 'social skin', in virtual environments apparel for avatars is simply referred to as 'skins', a revealingly embodied metaphor applied to a virtual identity. Unsurprisingly, companies who affiliate themselves with the metaverse narrative are eager to underscore the seemingly endless possibilities for the expression of identity that they provide. On the Meta Avatar Store, for example, it is claimed that avatars allow one to "be uniquely you" as well as "your authentic self" by "decking out your avatars with clothing from some of the world's leading brands" (Meta, 2022). And as Roblox CEO David Baszucki states, "we really believe everyone on our platform will ultimately be who they want to be and who you want to be" (McDowell, 2021).

The importance of virtual identities as a new market opportunity—and of video games and virtual worlds as a new space for brands to advertise and sell wares to young (proto-) consumers—can be witnessed from the multiple initiatives and collaborations that fashion companies both large and small have launched across and in almost all major virtual worlds. This development is in line with a longstanding history in which fashion companies have sought to capitalize on and advertise through various forms of entertainment and media, such as film, television, and photography. In light of this history, gaming platforms and metaversal environments can be understood simply as new marketing channels. However, the foray of such companies—from an industry that is known to celebrate traditional craftsmanship and the very material qualities of apparel, that is, distinctly embodied and material phenomena—into virtual spaces is clearly an entirely new and generally unfamiliar endeavour, and is characterized by much experimentation with various business models and different technologies, such as blockchains, NFTs, and cryptocurrencies. Games and virtual spaces, in other words, appear to pose different challenges and opportunities compared to previous forms of media.

Many fashion companies have sought to exploit the new possibilities that virtual spaces have to offer, often in the form of 'collaborations' that tend to involve the creation of fashion apparel and cosmetics that can be sold to gamers, or through the development of dedicated 'experiences' on major gaming platforms or virtual worlds. Although all manner of companies have sought out collaborations with gaming platforms and set up branded 'worlds' or advergames, it is fashion companies that have been the first to truly set up shop as businesses in metaverse environments, thus indicating the opportunity that both companies and people see for engaging with fashion on such platforms. That fashion companies are enthusiastically entering, or being drawn into, virtual worlds is particularly telling given their profound role in shaping embodiment and social identity, further highlighting the importance such worlds have in this regard, but also how virtual worlds are becoming more important for industries (and their consumers) that are often thought to be distant from the realm of 'tech'.

As noted in the previous chapter, the massively popular game Fortnite and its developer Epic are prime examples of the expansionary ambitions of metaverse platforms and companies. The Fortnite platform can be understood as one of the epicentres of contemporary pop culture and thus as a central reference point for the expression of identity, particularly for younger segments of the general population. This can be witnessed not only through the performances given by pop stars, but especially in the ways that Fortnite has established collaborations with numerous prominent brands, particularly ones associated with large media productions and personal lifestyle. Tellingly, almost all such collaborations involve the creation of skins and accessories for Fortnite avatars, which are subsequently purchased, frequently and repeatedly, through microtransactions (Sidorenko-Bautista et al., 2025). In-game appearances, one can easily say, are one of the platform's major revenue streams. Fortnite, including its most popular game mode Battle Royal, is free to play, but the majority of its revenue is generated through so-called 'battle pass' subscriptions, which grant access to, among other things, rewards and limited in-game items, such as weapons, apparel, and cosmetics. Another significant part of its revenue is generated through the Fortnite Shop where people can buy virtual goods, using Fortnite's currency V-Bucks, for their in-game characters or avatar by choosing from a plethora of items, accessories, and emotes that the platform has selected for this season's edition of their game modes. Both the Battle Pass and Fortnite Shop, in other words, revolve to a great extent around the appearance of the avatar and not, as one might expect, its in-game performance in Fortnite. Needless to say, this monetization of avatar appearance and the commodification of virtual identity through microtransactions is highly lucrative for gaming platforms (Li et al., 2020; Marder et al., 2019).

As already noted, Fortnite Battle Royal has 'seasons' that generally last around twelve to sixteen weeks, which each introduce new chapters to the game's storyline and are based on a new theme. Each successive season therefore allows for the introduction of new game features, maps, and events, but especially new and exclusive skins, cosmetics, and accessories, which are almost always branded (Carter et al., 2020). Perhaps unsurprisingly, Fortnite's seasonal model is similar to that of the fashion industry, which has since long let go of the notion that new clothing collections are related to the arrival of new (environmental) seasons. It is therefore no surprise that many prominent fashion companies have jumped on the opportunity to cater to the logic of infinite customizability in terms of identity, and to advertise and sell to gamers on Fortnite. American fashion company Ralph Lauren, to take one prominent example, has released multiple apparel and accessory collections involving updated and redesigned versions of their most-famous physical items, such as the Polo shirt via specially themed tournaments on Fortnite. Such collections, also known as 'drops', are generally introduced via livestreams on Twitch—a streaming platform that is central to gaming as a form of media for consumption—accompanied by claims such as, for example, "authentically expressing yourself is core to the player experience inside Fortnite" (Harrell, 2022). Tellingly, experience, embodiment, and 'authentic' identity are here not just unproblematically conflated but also situated as specifically virtual in character.

Due to the success and apparent long-term viability of virtual clothing lines, Ralph Lauren has gone further and created a separate on-platform space (called an 'island' on Fortnite) that involves a racing game and, of course, the release of a related collection of virtual racing apparel and cosmetics (McDowell, 2023). In other words, this is simultaneously an advergame and a sales channel. Other major apparel companies, such as Balenciaga, Moncler, Vans, and Nike, among others, have also released successful clothing lines and accessory collections on Fortnite, and in some cases also created Fortnite islands and related advergames promoting their brands. Similar initiatives can also be witnessed on other major gaming platforms, such as the collaborations of Benetton and H&M with Nintendo's Animal Crossing and of Burberry and Lacoste with Minecraft, not to mention the range of companies that have collaborated with Roblox, as will be discussed below.

It is certainly not only fashion companies that are seeking to affiliate their brands with the performance of identity through avatars on Fortnite. Prominent film franchises, anime series, pop stars, and car companies, to name but a few categories, are similarly involved in the sales of branded skins, accessories, and cosmetics. What is more, as noted in the previous chapter in the context of physical toy lines that emerge from gaming, the release of many of such virtual clothes is accompanied by the sale of similarly themed physical garments through conventional retail channels. Perhaps most interestingly, some of the new physical product lines were originally developed for virtual spaces only (Rinciari, 2024), thus again exemplifying how the influence of metaverse aesthetics is finding broader appeal beyond gaming platforms, as can also be witnessed from the recent and explicit use of Fortnite as a thematic backdrop at prominent runway shows as well as from the mimicking of emotes in goal celebrations by football players as well as on social media platforms. Last but not least, the outwards push of virtual aesthetics as an increasingly dominant cultural phenomenon can also be witnessed from the recent commercial success of the Minecraft movie, and the announcement of more upcoming game-based cinema productions.

It is clear that people on metaverse platforms such as Fortnite are engaging with virtual fashion in a manner similar to engaging with fashion in the physical world. Their avatars don and doff specific apparel in specific contexts and thus people negotiate social settings and perform their in-game identity through clothing. In the process, these actions and practices are commodified by both large apparel companies and technology companies. But for fashion companies there are opportunities beyond the brand collaborations in Fortnite described above, namely the creation of branded 'worlds' and experimentation with new business models and revenue streams, particularly on the more open-ended sandboxes. Although branded collaborations on platforms such as Fortnite emulate tried-and-tested approaches in marketing and sales, even more time and engagement with brands can be extracted from people on sandbox platforms such as Roblox, which are more focused on user-generated content and offer even higher degrees of customizability.

Prominent Italian fashion company Gucci, for instance, which already dipped its toes into the gaming world by designing accessories for popular games such as The Sims, Animal Crossing, and Pokémon Go, set up multiple temporary 'experiences' in virtual environments, such as Roblox, Zepeto, and The Sandbox, which generally lasted for a few weeks. As one of the blockchain-based platforms focusing on virtual real estate and cryptocurrencies, Gucci's 'experience' on The Sandbox initially revolved mostly around the sale of NFTs, advertising of past and current fashion items, and several 'play-to-know' advergames pertaining to the brand's product and historical heritage that allow players to buy or win items that avatars can wear. On Roblox, similarly, multiple worlds have been set up that allow people to try on and buy virtual items designed by Gucci—often virtual recreations of the brand's most famous items but also entirely new ones—as well as play advergames. The establishment of such virtual worlds and the sale of in-game apparel and accessories is, unlike NFTs, proving to be hugely successful. Due to the success of its temporary initiatives, which attracted tens of millions of people, Gucci has also created persistent or permanent 'experiences' as well with more elaborate modes of gamification, which allow players to 'unlock' new shopping areas and thus win the right to purchase more exclusive items. As a luxury brand seeking to maintain its 'aspirational' character, Gucci creates artificial scarcity not just through a tiered gamification strategy, but also seeks to draw in larger audiences for repeated visits by announcing limited-time 'drops' (Webster, 2022).

The hype around NFTs, which reached its peak in 2021–2022, appears, however, to have largely abated. Although NFTs initially appeared to present a highly lucrative alternative business model—enabling sales without the usual challenges of inventories and supply chains—the deployment of both cryptocurrencies and NFTs as a means for establishing ownership over virtual projects across gaming or metaverse platforms appears thus far to have failed. For instance, large virtual worlds such as Fortnite and Roblox already have a vibrant market economy based on their own on-platform currencies and ownership models, and although some occasionally seek to reinsert blockchain dynamics, these tend to be relatively marginal efforts. Furthermore, a telling example of the failure of cryptocurrencies and NFTs to create a foothold in virtual worlds, or even the fashion system more broadly, is the recent class-action lawsuit against Nike, which acquired the then highly hyped 'crypto-native' brand RTFKT in 2021, but this is now being shut down, causing significant losses to investors (Stempel, 2025). In other words, the assertion by NFT boosters as well as fashion companies that collectability, digital ownership, and scarcity-driven speculation would become an important part in the performance of virtual identities appears to have fallen on deaf ears. Moreover, one of the central claims circulating around blockchain technologies is that these provide a more participatory, collective, and decentralized approach to virtual worlds, thus serving as an important counterbalance to the centralized, top-down platforms operated by 'Big Tech' companies, such as Google and Meta (Herrman & Browning, 2021). The apparent failure of blockchain-based currencies and ownership models to be taken seriously in virtual worlds, as well as the lacklustre performance of blockchain-centric worlds such as Decentraland and The Sandbox, seem to indicate that the centralization by dominant companies is thus far continuing unabated and that it is therefore their business models that are still predominant.

Gucci is, to be sure, by no means the only apparel company that is exploring and experimenting with the sales and marketing opportunities that virtual worlds offer to young consumers in particular. Moreover, it's not just high-fashion companies that are being drawn into metaverse platforms, but also mid-level fashion companies, athletics brands, and cosmetics companies, which have not only set up virtual worlds or developed virtual items for games (sometimes, but certainly not always, blockchain-related NFT endeavours), but have also started to develop virtual try-on applications based on AR technology, set up virtual show rooms, and, in the case of beauty companies, created 'face tune' filters to virtually test out makeup. Some companies, such as Louis Vuitton and Balenciaga, have even released their own stand-alone games that are released through Google and Apple's app stores. AR applications in particular are interesting in terms of identity because they apply virtual layers to physical bodies, and thus have the potential to shape the experience and self-perception of embodiment in important ways.

As noted in the previous chapter, platforms such as Roblox rely predominantly on user-generated content and the labour, or 'playbour', of people who create not only the 'experiences' that attract players but also the plethora of items that circulate in the platform's virtual economy. A striking example of the centrality of both fashionable appearances and the creation of virtual spaces or games as well as in-game items by independent or amateur developers on Roblox is Dress to Impress, which is currently one of the most popular games on Roblox. Indeed, rather than merely an example of fashion and identity in a game, it is a 'fashion game' in itself. That is, the game's popularity demonstrates the central importance that people on Roblox place on virtual dressing and identity play.

The premise of Dress to Impress is deceptively simple. Players are presented with a theme and given a little over four minutes to create outfits that are related to this theme by selecting from a range of avatar customization options (e.g. hairstyle and makeup) as well as a range of virtual clothes in different styles that are available in the game's lobby. Once time is up, the assembled outfits are presented during a final runway show where other players rank each outfit on a five-star scale. All of this can be done for free. Compared to the monetization of content on social media platforms, where content creators indirectly earn revenue based on views and advertising or brand sponsorships and affiliate marketing, Dress to Impress creates direct revenue through in-game microtransactions. Players pay, for example, for premium access to a VIP area where more exclusive clothing and accessories can be selected, thus enhancing their chances of performing well on the runway contest that takes place once the game time is up. Such VIP items are highly recognizable and by definition scarce, which allows them to function as status symbols and encourages other players to pay for VIP access (Cappa, 2025). Other pay-to-access and pay-to-progress features are, for example, customizable makeup, which allows players to style their own makeup looks rather than being forced to select from pre-selected options, and additional runway poses. Similar to VIP items, customizable makeup and exclusive poses are perceived as giving players an edge over competitors during the final runway contest and increase one's likelihood of a higher score. Garnering hundreds of thousands of concurrent players, the game also appears to be attracting a more diverse set of age groups, thus indicating that Roblox is moving beyond its image as a platform predominantly for children (Roy, 2024). Dress to Impress thus symbolizes not just the centrality of the fashioning of virtual appearances on Roblox, but also of the platform's reliance on user-generated content as well as how Roblox seeks to commodify and monetize the various forms of production and consumption taking place there.

The 'freemium' model of platforms such as Fortnite and Roblox and their reliance on microtransactions by people in order to gain access to so-called loot boxes, random reward features, premium perks, and VIP access to restricted areas or items, as well as the multitude of apparel items, accessories, and dance moves available on their platform shops have not only proven enormously profitable but have also given rise to concerns about their gambling-like and therefore potentially addictive character (Brooks & Clark, 2019; Carter & Hardwick, 2025; Zendle & Cairns, 2019). Moreover, the logics of microtransactions and immersion in gaming—and how these can easily lead to overspending—bear a striking resemblance to that of microbets and betting 'flow' observed in ethnographies of gambling technologies and casinos (Schüll, 2012).

The performance of identity in metaverse platforms such as Fortnite and Roblox is certainly not only limited to the appearance of one's avatar. In-game performance (and success), levels of skill and knowledge, seniority and experience, as well as interactions with other players are similarly important. Nevertheless, the shift from physical to virtual clothing as a mode of expressing one's identity is another indication of the ways in which metaverse platforms are folding previously non-digital practices into their platform business models, and thus an important development in terms of the expression of personal identity. Moreover, the aesthetics of virtual identities and fashion are starting to loop back into conventional, physical apparel, thus exemplifying both the centrifugal and centripetal forces emanating from metaverse worlds.

Besides establishing itself as a central cultural phenomenon, Roblox is also becoming an increasingly central part to the so-called platform economy. Tellingly, Roblox seems to be 'maturing' in a manner reminiscent of the development of social media platforms. One example is that user-generated content from non-professionals is being supplemented with content by professional developers and game studios (Maguire, 2024). Additionally, some 'creators' are achieving levels of success that allow them to monetize their 'experiences' and thus become developers in their own right, such as has been the case for the creator of Dress to Impress. Another telling development in terms of the professionalization of content creation as well as Roblox's platformization along the lines of the 'Big Tech' model is that influencer marketing, sponsored content, and paid ads are starting to be introduced. For example, instead of creating their own branded worlds or selling items via the marketplace, companies are organizing 'pop up' shops and campaigns in already-popular and community-created 'experiences'. Such so-called brand integrations allow companies to experiment with and collect data on consumer engagement and sales metrics while capitalizing on the popularity of prominent creators (McDowell, 2024). In a similar vein, people on Roblox are starting to be rewarded with Robux and in-game 'power-ups' for watching advertisements, which is an additional development besides the paid partnerships for creators and the advertising techniques described in the previous chapter. All in all, these developments indicate how Roblox as well as similar virtual worlds are moving towards the 'Big Tech' business model, established by Meta and others, in which the actions and behaviours of people are collected and aggregated in the form of data that can subsequently be put to multiple profitable uses both on the platform itself as well as be sold to others for further commercial exploitation.

In virtual worlds such as Horizon Worlds, Fortnite, and Roblox, privately owned spaces operated by for-profit companies, monetization and commodification are a taken-for-granted reality, which means that everything that takes places within— whether productive 'playbour', identity performance, or gamified consumption— can be datafied and commodified (Joseph, 2021). Indeed, precisely because of the highly interactive nature of virtual world construction and gaming, virtual worlds can provide deep insights into, for example, all sorts of consumer behaviours, individual preferences, and cultural trends. This is one of the main reasons why fashion companies, but also companies from other sectors, are so eager to establish a presence on such platforms and to experiment with their various technical capabilities and the new business models that they engender. Behavioural data collected during interactions with 'experiences' as well as during the performance of identity offers insights into what people respond to and how but also on what they are likely to spend more time and money. In short, people and their on-platform actions and their subsequent data trail can be understood as a 'player commodity' (Nieborg, 2016). And, once again, the behavioural data of people on such platforms can be directly sold to data brokers, but can also be used for algorithmic profiling, generating novel business models, creating new products and services, and for increasing the value of existing assets, not least of which the platform itself. Virtual worlds are thus positioning themselves as spaces not only for social interaction and for the performance of identity but also for new forms of labour and consumption. In the meantime, such virtual worlds are also creating new opportunities for commodification and datafication.

## 4.5  Virtualizing Bodies

Before the current rise of gaming platforms as social spaces, it was social media companies, such as Meta and its subsidiaries Facebook and Instagram, which established themselves with increasing success as new arenas for social interaction and self-expression, especially in supposedly more free and creative ways. The advent of virtual worlds can be seen as a continuation as well as intensification of these endeavours, particularly in how they seek to extend the notion that virtual identities are becoming increasingly important if not more important than our physical bodies. Simultaneously, virtual worlds are drawing in more and more professional as well as leisure practices while commodifying and datafying these in the process. However, one of the core assertions of this book is that 'metaversification' is not only a process of pulling activities and practices that were previously carried out elsewhere into virtual platforms, but also that the logics of virtualization and datafication are being increasingly pushed outwards, into our private and public environments as well as our bodies.

An obvious starting point for investigating this emergent process is the development and use of AR technologies, which, in contrast to VR technologies that effectively cut one's sensory perception off from the surrounding environment and people, instead seek to maintain a balance between virtual elements and environmental ones. Or in other words, AR allows us to extend our physical reality, and through this blending of digital elements with the physical environment, "reality becomes a hybrid" (Bolter et al., 2021, xix). Unlike VR technologies that are, at least for now, generally envisioned for inside use in somewhat private spaces, AR technologies are explicitly envisioned as being used while navigating through public spaces, which can be both indoors and outdoors. That is, they are worn and used in environments in which one is likely to have embodied interactions with others. This has important implications in terms of datafication and virtualization as AR technologies directly shape our perception of the physical environment as well as our bodies and those of others through the insertion of virtual elements. It is perhaps not for nothing that some prominent developers of AR technologies refer to these with visual metaphors such as 'lenses' and are increasingly seeking to integrate AR into so-called 'smart' glasses.

AR, like VR, is more usefully understood as an umbrella term for various technologies that share a common feature, namely that they overlay virtual and non-virtual aspects, but it is also important to note that these varieties of AR nonetheless have different uses and implications. On the one hand, Microsoft's HoloLens, for example, is a headset used mostly in professional settings (particularly by the police and military, but also in various industrial contexts) and Apple's Vision Pro appears mostly geared towards leisure uses. On the other hand, Pokémon Go is a game app used via smartphones, and the functionalities of the 'smart glasses' developed by Meta and EssilorLuxottica are, for now, somewhat akin to those of smartwatches. In short, although AR is often envisioned as involving the donning of headsets or glasses, an already very common use of AR is via smartphones, which are already ubiquitous and in widespread use. Despite such diversity in form and function, AR technologies extend the logics of virtuality, and as we will see below also those of datafication, outwards from the realm of virtual worlds into the environment and thus provide a bridge between virtual identities and physical bodies.

One AR application that is particularly relevant to the performance of identity, and therefore also to embodiment, are so-called face filters or face tuning apps, which are generally used to edit or augment selfies that are taken with smartphones and uploaded to social media platforms, but can also be used to change one's appearance on video calls and livestreams. Like the narratives that circulate around the performance of identity on social media platforms and virtual worlds, AR filters are frequently framed positively by companies who develop such technologies as facilitating creative experimentation with the presentation and performance of one's self-identity while trying out different looks. Unsurprisingly, as industries deeply invested in appearances, fashion and cosmetics companies in particular have sought to exploit this new media form because they allow potential customers to virtually try on their products, whether this is the superimposition of a shoe on a foot or the overlaying of lipstick on lips. Although face filters can also be used to superimpose fantastical or unrealistic elements onto one's face, the most widespread use of AR filters, however, is the 'beautification' of selfies that are shared on social media platforms, which generally consists of the smoothing, lightening, and diminution or exaggeration of certain facial features. That is to say, beautification filters are used to make apparently realistic, or at least plausibly realistic, adjustments to one's face. However, due to the often normative and idealized uses of face filters—i.e. adjustments that often remain somewhat invisible to others who are exposed to such 'augmented' representations of bodies—concerns have been raised about AR filters for reproducing unrealistic beauty standards and thus leading to distortions in bodily self-perception and a subsequent reduction in psychological wellbeing (Miller & McIntyre, 2022; Szambolics et al., 2023). In this sense AR technologies can profoundly reshape, for better or for worse, how we see the bodies of others as well as those of ourselves.

Although originally often associated with the social media company Snapchat (now known as Snap), AR filters have increasingly been taken up across other social media platforms, such as Instagram and Tiktok, but also by live streamers on YouTube and Twitch who sometimes employ virtual avatars to represent themselves. AR filters and their role in the representation and mediation of identities can be understood as falling somewhere between the operations of established social media platforms and the employment of avatar identities in virtual spaces. It is for this reason that some have called this intersection of smartphone-enabled filters and virtualized identities an example of the "actually existing metaverse" because AR filters appear to function as a halfway point for companies such as Meta in their goal towards drawing people into their specific world (Bozzi, 2024). Companies such as Meta and Snap have also sought to position AR filters as a new and valuable type of 'content' by drawing in creators and developers into their respective Snap AR and Spark AR studios for the building of ostensibly augmented 'experiences'. It is perhaps also telling that Snap is one of the few social media companies besides Meta that has heavily invested in the development of AR glasses. Crucially, AR filters and the various forms of aesthetic and identity experimentation they enable can be seen as an important means for engaging and habituating people into new emerging 'Big Tech' infrastructures such as the metaverse.

AR technologies, not just face filters but also the wide range of other AR applications, function as a bridge between the logics of virtual spaces or worlds and the more familiar experience of moving through and living in the physical world. Needless to say, the information that can be collected about faces and other body parts through AR filters is yet another opportunity for datafication (Eugeni, 2024). AR filters thus present one important example of the virtualization of bodies and identities, and of how such representations can circulate on social media platforms but also loop back into our lived experience and self-perception. In an important sense, however, AR technologies and the virtualization of identity that they enable remain firmly rooted in our physical or embodied reality.

Another instance in which the virtualization of bodies is relevant is the creation of avatars used to play games or navigate in virtual environments as well as the virtual characters that are encountered and interacted with in games (also known as Non-Playable Characters or NPCs). As already noted several times before, a virtual rendition called the avatar is posited as representing ourselves in gaming environments or virtual worlds. In one influential version of this, avatars are:

> living 3D representations of you, your expressions, your gestures that are going to make interactions much richer than anything that's possible online today. You'll probably have a photo realistic avatar for work, a stylized one for hanging out and maybe even a fantasy one for gaming. (Meta, 2021c)

Zuckerberg here portrays the avatar as a 'living representation' of oneself that can be 'photorealistic', suggesting that the avatar is akin to our embodied reality and capable of reproducing, or at least representing, our abilities to communicate with others in a straightforward manner. This definition of the avatar as a real-time 3D entity is relatively new, however, and is closely tied both to the emergence of virtual environments that are not specifically games and to the technical capacity to graphically render convincingly lifelike virtual representations. In earlier video games, one's avatar was usually the main character of the game and could barely, if at all be, customized, and due to the state of graphics rendering capabilities, generally quite cartoon-like. With the advent of role-playing games and massive, multi-player online games, as well as concurrent developments in graphics processing and game physics, the options for avatar customization and photorealism dramatically increased. However, it is only with the emergence of virtual worlds such as Second Life—and the subsequent idea that virtual spaces can mirror reality—that avatar creation geared towards the reproduction of actual people becomes more prevalent. Nevertheless, it should also be remembered that although many people model their avatars to some extent after their physical selves, or at least remain quite close to this, plenty of people do not seek to model their avatars after anything that resembles reality as we generally know it. Moreover, games that overtly seek to avoid such forms of realism continue to be created and remain popular. Indeed, Roblox and Minecraft provide avatars that appear to emulate Lego toys more than human bodies.

As noted in the previous chapter, at the centre of the production of 'reality' in virtual environments is the game engine, which enables the 3D modelling, physics simulation, and real-time rendering that together make virtual environments appear as realistic or lifelike worlds. Due to their ability to render virtual images in increasingly realistic ways, game engines have become central to industries that similarly seek to recreate and represent imaginary worlds, such as film and cinema production as well as architecture, construction, and design. Moreover, due to their ability to virtually recreate complex environments, game engines have also become central to the development of so-called 'digital twins', which are employed to monitor and operate a variety of urban, industrial, or infrastructural systems and processes (see previous chapter). Much of the initial creation of video games tends to focus on the construction of virtual spaces that serve as the setting for a game, and the use of game engines in cinema, for example, is generally employed to provide realistic scenes or backgrounds for actors. In film and cinema, game engines are employed to generate 'special effects' or Computer-Generated Imagery (CGI) to create scenes in cinema that would otherwise be impossible (or prohibitively expensive) to recreate. In addition to creating spaces or scenes, however, it is also necessary to populate games and film productions with characters with which gamers or actors interact in a seemingly natural or realistic manner. In other words, game engines are also employed for the creation of human-like characters, also known as 'digital humans' or 'virtual humans', with which we, via our avatars, are expected to interact in virtual environments.

The reproduction of photorealistic environments, and perhaps physical realism more broadly, has been a guiding ideology in graphic game production since its early days and is still frequently upheld as a benchmark, particularly in large and expensive 'blockbuster' games (Phillips, 2020). Although computer-generated images of physical environments have been mostly indistinguishable from 'real' photographic images for some time now—at least to the non-expert eye—photorealistic representations of humans and their actions have only recently become difficult to distinguish from non-computer-generated imagery. Nevertheless, relatively realistic, let alone real-time, representations of humans in games and virtual worlds—and particularly natural-like movements of facial features—continue to be considered one of the most technically difficult endeavours to accomplish (Sito, 2013). This is in large part due to the multiple complexities involved in rendering convincing-looking human skin and how its texture and partial opacity interact with light, but also because bodily movements are enormously complicated to simulate in a convincing, that is, non-uncanny manner. Due to the significant amounts of labour involved in the usual way of modelling virtual humans via polygon meshes and wireframes, and their often-unsatisfactory outcomes, the gaming and entertainment industries have turned to techniques such as volumetric and 3D scanning to capture bodily movements and bodies themselves. In addition to the photogrammetry techniques used to capture objects and environments, these capturing techniques are used to construct large 'asset libraries' that are subsequently used to simplify and automate the modelling, texturing, and animating of virtual humans. It is hence quite apt to say that "reality is being captured to create 3D worlds" (Chia et al., 2023, 15).

Epic Games, as also noted in the previous chapter, is one of the largest companies involved in the creation of such asset libraries, which is an invaluable resource for its game engine Unreal, itself one of the two dominant tools, the other being Unity, used for the automation of 3D worldbuilding across multiple industries. Recent corporate acquisitions by Epic involve photogrammetry studios such as Quixel and Capturing Reality as well as startups such as 3Lateral and Cubic Motion that specialize in the capturing of humans for the purpose of their virtual recreation. Given its expertise in the creation of games and virtual worlds as well as the developments of the tools to create such worlds, it is to be expected that Epic would also position itself as a dominant player in the production of virtual humans and avatars that are foreseen as populating the metaverse.

Epic's MetaHuman Creator is a tool that is marketed as giving "anyone the power to create, animate, and use highly realistic digital human characters in any way imaginable" (Epic, n.d.). This free, cloud-based app is integrated with the Unreal Engine and allows one to select from a range of diverse presets with which one can swiftly craft realistic human faces and body movements as well as facial animations. These virtual humans, or 'metahumans' as Epic prefers to call them, are fully 'rigged', meaning ready for use in animation and enabled for real-time motion capture, and can be inserted into various video games, virtual worlds, or film and cinema productions. Their individual characteristics (hairstyles, facial features, clothing, makeup) can be further tailored to meet the precise needs of each type of production. It is not just the creation of digital humans but also their animation that MetaHuman Creator seeks to simplify and automate. Through partnerships with companies that specialize in using computer vision and AI for the automation of facial capturing and animation, such as Faceware and Digital Domain, Epic seeks to create a direct connection between human performers and their virtual counterparts that no longer relies on human intermediaries (i.e. animators) and instead only on technology and automation (McKim, 2022). In other words, Epic's slogan of "high-fidelity digital humans made easy" involves not only the capturing of reality for the creation of 3D worlds but also, paradoxically, the replacement of human labour for the creation of virtual humans. Epic's virtual humans, in short, are intended to populate currently emergent virtual environments, above all the 'metaverse', but remain tightly tethered to its Unreal

Engine. That is, their existence is, and to some extent our virtual worlds are becoming, increasingly reliant on the "reality capture infrastructure of engine tools" such as those made by Epic (Chia et al., 2023).

The photogrammetry efforts by Epic to scan human bodies are claimed to represent "all possible appearances [...] that people can take" (Epic, 2021). In other words, Epic's database is claimed to be 'diverse' because it contains 'any human appearance', or at least a statistically distributed representation of this. This 'datalogical' conceptualization of diversity as based on scans of human appearances is, almost literally, only skin deep and can be criticized as representing a form of 'diversity surfing' (Freedman, 2020). It is also problematic in the sense that it reduces the complexity of identity to decisions made by an ostensibly objective statistical model, leaving out any questions about the construction of this model and its underpinning logics. In other words, "Meta- Human positions diversity as preexisting in the statistical compression of its photogrammetric database, preceding the need to classify or even recognise race" (Chia, 2025, 175). Moreover, Epic's conceptualization of diversity and its repackaging of this through photogrammetric scans and data is predominantly asset driven, that is, they are intended for commercial use and sale. An obvious sense in which Epic's pre-made virtual humans are an asset class is their intended use and indeed re-use in games, film and cinema production, and social media. That is to say, rather than only being a tradeable commodity that can be sold to others, virtual humans are a resource that can continuously generate future income. However, in contrast to the assetization of virtual objects and 'land' discussed in the previous chapter, which have thus far been relatively unsuccessful, virtual humans created through tools us as Meta Human Creator are starting to become 'infrastructural assets' (Chia, 2025, 186–187) in a wider system of production and labour in which game engines and the companies that create and operate them are becoming central to other industries, such as entertainment, architecture, and design, among others (Jungherr & Schlarb, 2022). One particularly telling example of the application of virtual humans, beyond that of non-playable characters in games, is the emergence of close-to photorealistic chatbots. In short, virtual humans as faces for AI in what might be called a form of 'artificial sociality' (Natale & Depounti, 2024). The integration of the AI language capabilities with human-like interfaces is thought to enable better, i.e. more humanlike services (which are also cheaper and more scalable as well as, once again, infinitely customizable, so perhaps not so humanlike after all), in areas such as sales assistance, customer service, and corporate training, or even provide an alternative form of companionship and friendship (Hitti, 2020; Seymour et al., 2023). A perhaps more familiar example of the application of virtual humans as a 'digital workforce' (Ong, 2021) is on social media platforms, where virtual influencers such as Miquela Sousa and Lu do Magalu, have amassed millions of followers while sharing their 'travels' and 'lifestyles', thus inserting themselves as viable additions or perhaps alternatives to the commodification of human 'experiences' on such platforms. In other words, the logics of virtualization are, once again, being pushed outwards into new spaces both physical and digital.

## 4.6  Datafying Bodies

As examined in the previous chapter, AR and VR technologies are an important means for extending the logics of virtualization and datafication by 'Big Tech' companies into our environments. In addition to enabling the collection of data about the spaces in and through which we live and move, such devices can also collect reams of data about us, and particularly our bodies and behaviours. And although smartphones and various wearables are already used for datafying our actions and behaviours, indeed sometimes voluntarily, AR and VR headsets promise to collect even larger and potentially more insightful types of data about us.

Unsurprisingly, companies such as Meta and other headset manufacturers rarely, if at all, talk about such data, at least not in public settings, although they do occasionally and rather superficially gesture towards concerns about the safety and privacy of individual 'users'. During Meta's Connect 2021 conference, for example, the only mentions of data are made in two specific contexts. One is a passing comment that asserts that it is important to be "transparent" about "what data is collected, and how that data is used over time", while the other mention of data is during a particularly revealing demonstration in which a woman wearing AR glasses is shown navigating a domestic space. While walking through this home, the glasses map in precise detail everything to which she turns her head. This mapping, it is explained, allows Meta to create an "index" of "every single object", and by tracking her eye movements, to figure out what she is "interested in" and offer "proactive assistance". Although this setting is by any definition an extremely datafied one—and allusions are made to the tracking of bodily movements and actions in order to anticipate a person's intentions and desires—Meta is only explicit about how datafication is done to objects. However, the opportunities for commodification and advertising even on the basis of data only about objects—not to mention data on personal behaviours and desires—are countless.

It has already been observed that Meta's conceptualization of the 'Metaverse' is surprisingly full of bodily actions and pleasures. Above all, the metaverse and the mixed-reality technologies developed by Meta are portrayed as being informed by human actions, answering to human needs, and controlled by humans. However, the interfaces that are being operated by 'users' also open up their bodies to being read by these technologies. Meta's interest in such bodily sensing technologies can be seen from its acquisitions of startups that develop technologies for mapping, and interacting with bodies. In 2020, Meta acquired Lemnis Technologies, a headset manufacturer that specifically focuses on eye-tracking technology. But Meta also has ambitions that go well beyond such haptic and ocular interfaces and into 'neural interfaces' that will purportedly allow for the translation of neuromotor signals into digital commands with which one can operate devices both physical and virtual. CTRL-Labs, for example, was acquired in 2019 and focuses on developing wrist-bands that translate neuromuscular signals into machine-readable commands. In combination with predictive algorithms that calibrate themselves to our individual desires and preferences, as well as contextual AI, which understands our environment and predict our orientations towards it, such neural interfaces will provide a seamless and frictionless pathway between our bodies and technology, or as Zuckerberg puts it, we will "be able to send a text message just by thinking about moving your fingers" (Meta, 2021c). In a related note, Elon Musk—not commonly associated with the metaverse but certainly a figure who exemplifies some of the most hubristic if not totalitarian tendencies of 'Big Tech'—is similarly seeking to establish such a form of disintermediated communication with his company Neuralink, which develops brain-computer interfaces that Musk himself describes as "a Fitbit in your skull" (Regalado, 2020). The suggested relationship between such interfaces and processes of datafication and (self-)quantification are unlikely to be accidental. Once again, although the body of the 'user' is framed as the one being in control, what is left unsaid, and presumed as normalized, is that using such thought-interpreting devices will allow one's data to constantly be transmitted and collected.

In terms of devices that are already in widespread use, both VR and AR are fundamentally reliant on spatial and bodily data, because these provide the foundation upon which virtual layers and environments can be created that can not only be perceived but also manipulated by people wearing such technologies. One of the primary concerns therefore is that VR and AR technologies can collect bodily data on a far more granular level. As noted by Bailenson, for example, such technologies can "record 18 types of movements across the head and hands", and "spending 20 min in a VR simulation leaves just under 2 million unique recordings of body language" (2018, 905). Such data on bodily movements, however, can be diagnostic of personal identity, medical conditions, and mental states and thus fall under the definition of biometric data (Miller et al., 2020; Pfeuffer et al., 2019). In addition to data on bodily behaviours, VR and AR technologies are also capable of eye-tracking, gaze analysis, voice and facial recognition, and potentially also electrical muscle activity, respiratory or heart rate, and galvanic skin response (Cross & Coby, 2023). Even relatively small amounts of such granular data can be highly revealing—ranging from age, gender, height to (dis)ability and motor-response times—and as accurate as fingerprint data for identifying individuals (Nair et al., 2023). Such concerns notwithstanding, Meta has recently announced it will start collecting anonymized data via Quest headsets for "building better experiences" (Orland, 2024).

This type of datafication of bodies opens people up to multiple potential vulnerabilities and harms, such as corporate surveillance as well as sousveillance. This is certainly not only the case for commercial consumers of mixed-reality technologies, but above all for workers in, for example, manufacturing, warehousing, and logistics who are already being required to use such technologies and whose labour has already since long been subjected to processes of quantification, automated decision-making, 'nudging', and remote surveillance through so-called assistive technologies that are introduced under the guise of optimization, productivity, and efficiency (Bonini & Treré, 2024; Delfanti, 2021). As Evans et al. assert (2022, 29), virtual worlds "can act as a fully integrated dataspace" that enables unprecedented levels of data collection, algorithmic recommendation, and commodification, but this is equally true for environments in which metaverse technologies are worn, particularly when these technologies are increasingly becoming a mass-market phenomenon. Moreover, as noted by Harley (2020), mixed-reality technologies are frequently presented by 'Big Tech' companies via colonial metaphors that betray how the 'metaverse' is understood as a 'new frontier' that is waiting to be discovered and, by implication, appropriated and occupied, just as the internet was in the 1990s (Flichy, 2007). As such, metaverse datafication presents both an extension and intensification of the existing business models of 'Big Tech' companies—perhaps above all of Meta—which consist of a so-called ecosystem in which the new and more intimate data can be collected about our bodies and our selves, and subsequently put to new uses.

## 4.7   Colonizing Bodies

As noted at the end of the previous chapter, the distinction between the current furore around generative AI and the hype around the metaverse that preceded this are perhaps not as disconnected as might initially appear. The datafication of public spaces that metaverse technologies such as VR and AR can contribute to is an important means for AI systems to move beyond their current predominantly text- and image-based understanding of the world. To this end, the collection of data that will allow AI systems to learn about spatiality and grasp the physics of activities and interactions in space is already being undertaken. Moreover, multiple companies have started to release large spatial models that enable AI systems that are integrated into robots—yet another frontier to be conquered by 'Big Tech'—to move through and interact with physical environments. Spatiality is, of course, only one side of the coin. Corporeality, and particularly how bodies move and interact as well as navigate environments is an equally important step for robots. But although AI systems have learned quite a bit about 'talking the talk', they still have some way to go in terms of 'walking the walk'.

One of the things this chapter has shown is that metaverse technologies can collect data about our bodies as well as how our bodies interact with and move through physical spaces. Such multi-modal sets, that is, combining multiple types of data on diverse phenomena, are important for teaching AI systems to get a better sense of 'how the world works', and particularly how humans act, operate, and live in this world. Or perhaps more accurately, 'how humans work'. To this end so-called 'large behavioural models' are starting to be released, which, rather than focusing on just textual information, also analyse behavioural patterns as well as visual, auditory, and other sensory data. In addition to speaking in a seemingly human manner, such models can, it is claimed, interact with the world in a more 'natural' and 'human-like' manner. Importantly, interactions with the world also include virtual worlds, as large behavioural models can be used to simulate human behaviours in gaming worlds. However, 'real world' interactions are equally relevant as such models can be applied to, for example, autonomous vehicles or assistive robots.

Robotics companies have started to become more and more keen on under-scoring how robots are increasingly capable of completing particularly everyday tasks. Figure, a robotics company that is collaborating with OpenAI, a notorious frontrunner in the development of Large Language Models, released a video in early 2024 that demonstrated how its flagship robotics project—Figure 1—is capable of identifying and handling delicate objects such as apples and tableware while also engaging in some very human-like speech patterns, such as occasional hesitancies and stutters. Figure's CEO, Brett Adcock, claimed in a 'master plan' published on the company's website that Figure 1 will "create a better life for future generations" and allow "us to live happier, more purposeful lives" (Adcock, 2022). Although dish washing is impressive, this demonstration was held in a controlled lab environment. A seemingly more impressive demonstration was Tesla's recent introduction of several 'Optimus' robots during a recent car event in which the robots mingled with the crowd, engaged in conversations, served drinks, and even appeared to dance. With familiar humility, Tesla's CEO Elon Musk has suggested that these 'Teslabots' will allow for "a future where there is no poverty" and will be a "fundamental transformation of civilization as we know it" (Shakir, 2024). Ironically, however, the name Optimus is inspired by the Transformers media franchise in which two robot factions are locked in a perpetual war and human characters are generally relegated to supporting roles at best.

# References

Adcock, B. (2022, May 20). *Master plan*. FigureAI. Retrieved March 15, 2025, from https://www.figure.ai/master-plan

Andreessen, M. (2023, October 16). *The techno-optimist manifesto*. Andreessen Horowitz. Retrieved March 15, 2025, from https://a16z.com/the-techno-optimist-manifesto/

Bailenson, J. (2018). Protecting nonverbal data tracked in virtual reality. *JAMA Pediatrics, 172*(10), 905–906. https://doi.org/10.1001/jamapediatrics.2018.1909

Barcella, D., Benecchi, E., Fomasi, M., & Balbi, G. (2025). Stop calling the Web the Internet: CERN's tactics to fight 'brand confusion' in the mid-1990s. *Convergence, 0*(0). https://doi.org/10.1177/13548565241308749

Barlow, J. P. (1996, February 8). *A declaration of the independence of cyberspace*. Retrieved March 15, 2025, from https://www.eff.org/cyberspace-independence

Baym, N. K. (2010). *Personal Connections in the Digital Age*. Polity Press.

Bolter, J. D., Engberg, M., & MacIntyre, B. (2021). *Reality media: Augmented and virtual reality*. MIT Press.

Bonini, T., & Treré, E. (2024). *Algorithms of resistance: The everyday fight against platform power*. MIT Press.

Brooks, G. A., & Clark, L. (2019). Associations between loot box use, problematic gaming and gambling, and gambling-related cognitions. *Addictive Behaviors, 96*(96), 26–34. https://doi.org/10.1016/j.addbeh.2019.04.009

Boellstorff, T. (2015). *Coming of age in second life: An anthropologist explores the virtually human second edition*. Princeton University Press.

Bollmer, G. (2017). Empathy machines. *Media International Australia, 165*(1), 63–76. https://doi.org/10.1177/1329878X17726794

Boyd, D. (2014). *It's complicated: The social lives of networked teens*. Yale University Press.

Bozzi, N. (2024). Meta's artistic turn: AR face filters, platform art, and the actually existing metaverse. *Information, Communication & Society, 28*(5), 832–851. https://doi.org/10.1080/1369118X.2024.2427116

Cappa, M.L. (2025) *From Web 2.0 to the Metaverse: Analyzing the evolution of platform-based business models and the creator economy* [Master's Thesis, Politecnico di Milano]. https://www.politesi.polimi.it/handle/10589/236207

Carter, M., & Egliston, B. (2024). *Fantasies of virtual reality: Untangling fiction, fact, and threat.* MIT Press.

Carter, M., Moore, K., Mavoa, J., Horst, H., & Gaspard, L. (2020). Situating the appeal of Fortnite within children's changing play cultures. *Games and Culture, 15*(4), 453–471. https://doi.org/10.1177/1555412020913771

Carter, M., & Hardwick, T. (2025, March 23). "Literally just child gambling": What kids say about Roblox, lootboxes and money in online games. *The Conversation.* https://theconversation.com/literally-just-child-gambling-what-kids-say-about-roblox-lootboxes-and-money-in-online-games-250387

Chia, A. (2025). Photogrammetric race-making in the MetaHuman Creator. In S. Mitchell, C. Perry, S. Redmond, & L. Torre (Eds.), *The screens of virtual production: What is real?* (pp. 174–193). Routledge. https://doi.org/10.4324/9781003463139-15

Chia, A., Malazita, J. W., Young, C. J., Nieborg, D. B., Joseph, D. J., & Gantt, M. D. (2023). The engine is the message: Videogame infrastructure and the future of digital platforms. *AoIR Selected Papers of Internet Research, 2022.* https://doi.org/10.5210/spir.v2022i0.12954

Costa Pinto, D., Reale, G., Segabinazzi, R., & Vargas Rossi, C. A. (2015). Online identity construction: How gamers redefine their identity in experiential communities. *Journal of Consumer Behaviour, 14*(6), 399–409. https://doi.org/10.1002/cb.1556

Cross, R. J., & Coby, E. (2023, December 16). VR risks for kids and teens. *PIRG.* https://pirg.org/edfund/resources/vr-risks-for-kids/

Delfanti, A. (2021). *The warehouse.* Pluto Press.

Epic Games. (n.d.). *Unreal engine—MetaHuman.* Retrieved March 15, 2025, from https://www.unrealengine.com/en-US/metahuman

Epic Games. (2021). *The rise of real-time digital humans on The Pulse.* Retrieved March 15, 2025, from https://www.unrealengine.com/en-US/blog/the-rise-of-real-time-digital-humans-on-the-pulse

Eugeni, R. (2024). A scanner darkly: Augmented reality face filters as algorithmic images. *Visual Communication, 23*(3), 498–512. https://doi.org/10.1177/14703572241235286

Evans, L., Frith, J., & Saker, M. (2022). *From microverse to metaverse: Modelling the future through today's virtual worlds.* Emerald Group Publishing.

Flichy, P. (2007). *The internet imaginaire.* MIT Press.

Freedman, E. (2020). *The persistence of code in game engine culture.* Routledge.

Gebru, T., & Torres, É. P. (2024). The TESCREAL bundle: Eugenics and the promise of Utopia through artificial general intelligence. *First Monday, 29*(4). https://doi.org/10.5210/fm.v29i4.13636

Harley, D. (2020). Virtual Bodies Inc., Framing corporate mediations of bodies in VR. *Public, 30,* 250–259. https://doi.org/10.1386/public_00019_7

Harrell, A. (2022, November 2). How metaverse fashion is fueling the future. *Sourcing Journal.* https://sourcingjournal.com/topics/technology/metaverse-digital-fashion-ralph-lauren-roblox-fortnite-umbro-parsons-stageverse-avatar-385378/

Hayles, N. K. (1999). *How we became Posthuman: Virtual bodies in cybernetics.* University of Chicago Press.

Heft-Luthy, S. (2019, August 28). The myth of the "empathy machine." *The Outline.* https://theoutline.com/post/7885/virtual-reality-empathy-machine

Herrman, J., & Browning, K. (2021, July 10). Are we in the Metaverse yet? *The New York Times.* https://www.nytimes.com/2021/07/10/style/metaverse-virtual-worlds.html

Hitti, N. (2020, January 15). Samsung's artificial Neon humans are "a new kind of life." *Dezeen.* https://www.dezeen.com/2020/01/15/samsung-neon-star-labs-artificial-humans/

Joseph, D. (2021). Battle pass capitalism. *Journal of Consumer Culture, 21*(1), 68–83. https://doi.org/10.1177/1469540521993930

Jungherr, A., & Schlarb, D. B. (2022). The extended reach of game engine companies: How companies like Epic Games and Unity Technologies provide platforms for extended reality applications and the metaverse. *Social Media + Society, 8*(2), 1–12 https://doi.org/10.1177/2056305122117 07641

Kalpokas, I., & Kalpokienė, J. (2024). I VR therefore I am: Toxic binary thinking in visions of the metaverse. *Information, Communication & Society, 28*(5), 910–925. https://doi.org/10.1080/1369118X.2024.2427119

Li, L., Freeman, G., & Wohn, D. Y. (2020). Power in skin: The interplay of self-presentation, tactical play, and spending in Fortnite. In *Proceedings of the annual symposium on computer-human interaction in play, CHI PLAY '20* (pp. 71–80). ACM. https://doi.org/10.1145/3410404.3414262

Lupinacci, L. (2022). Reclaiming the experience: social media, the "metaverse" and extractive imaginaries of experiential enhancement. *AoIR Selected Papers of Internet Research, 2022.* https://doi.org/10.5210/spir.v2022i0.13045

Maguire, L. (2024, April 10). The Gang: The startup that builds Roblox worlds for Gucci and Givenchy. *Vogue Business.* https://www.voguebusiness.com/story/technology/the-gang-the-startup-that-builds-roblox-worlds-for-gucci-and-givenchy

Marder, B., Gattig, D., Collins, E., Pitt, L., Kietzmann, J., & Erz, A. (2019). The Avatar's new clothes: Understanding why players purchase non-functional items in free-to-play games. *Computers in Human Behavior, 91,* 72–83. https://doi.org/10.1016/j.chb.2018.09.006

McDowell, M. (2021, May 17). Inside Gucci and Roblox's new virtual world. *Vogue Business.* https://www.voguebusiness.com/technology/inside-gucci-and-robloxs-new-virtual-world

McDowell, M. (2023, August 1). Ralph Lauren goes bigger on Fortnite with island and phygital boots. *Vogue Business.* https://www.voguebusiness.com/technology/ralph-lauren-goes-bigger-on-fortnite-with-island-and-phygital-boots

McDowell, M. (2024, February 27). Influencer marketing has arrived on Roblox. *Vogue Business.* https://www.voguebusiness.com/story/technology/influencer-marketing-has-arrived-on-roblox

McKim, J. (2022, April 11). Animation without animators: From motion capture to MetaHumans. *Animationstudies 2.0.* https://blog.animationstudies.org/?p=4426

Messeri, L. (2024). *In the land of the unreal: Virtual and other realities in Los Angeles.* Duke University Press.

Meta (n.d.). *VR for good: Virtual reality storytelling focused on social impact.* Meta. Retrieved March 15, 2025, from https://www.meta.com/en-gb/community/vr-for-good/?utm_source=about.meta.com&utm_medium=redirect

Meta. (2021a). *Founder's letter, 2021.* Retrieved March 15, 2025, from https://about.fb.com/news/2021/10/founders-letter/

Meta. (2021b, October 28). Connect 2021: Our vision for the Metaverse. *Tech at Meta.* Retrieved March 15, 2025, from https://tech.facebook.com/reality-labs/2021/10/connect-2021-our-vision-for-the-metaverse/

Meta. (2021c, October 28). The Metaverse and how we'll build it together—Connect 2021. *YouTube* (video). https://www.youtube.com/watch?v=Uvufun6xer8&ab_channel=Meta

Meta. (2022, June 20). *Introducing the meta avatars store.* Meta. Retrieved March 15, 2025, from https://about.fb.com/news/2022/06/introducing-the-meta-avatars-store/

Miller, D. (2012). *Consumption and its consequences.* Polity.

Miller, L. A., & McIntyre, J. (2022). From surgery to Cyborgs: A thematic analysis of popular media commentary on Instagram filters. *Feminist Media Studies, 23*(7), 3615–3631. https://doi.org/10.1080/14680777.2022.2129414

Miller, M. R., Herrera, F., Jun, H., Landay, J. A., & Bailenson, J. N. (2020). Personal identifiability of user tracking data during observation of 360-degree VR video. *Scientific Reports, 10*(1). https://doi.org/10.1038/s41598-020-74486-y

Morozov, E. (2013). *To save everything, click here: The folly of technological solutionism.* PublicAffairs.

Nair, V., Guo, W., Justus Mattern, B., Aachen, R., Wang, R., O'Brien, J., Rosenberg, L., Mattern, J., & Song, D. (2023). Unique identification of 50,000+ virtual reality users from Head & Hand motion data. In *Proceedings of the 32nd USENIX conference on security symposium SEC '23* (pp. 895–910). USENIX Association. https://www.usenix.org/conference/usenixsecurity23/presentation/nair-identification

Natale, S., & Depounti, I. (2024). Artificial sociality. *Human-Machine Communication, 7*, 83–98. https://doi.org/10.30658/hmc.7.5

Newton, C. (2021, July 22). Mark Zuckerberg is betting Facebook's future on the Metaverse. *The Verge.* https://www.theverge.com/22588022/mark-zuckerberg-facebook-ceo-metaverse-interview

Nieborg, D. B. (2016). Free-to-play games and app advertising: The rise of the player commodity. In J. F. Hamilton, R. Bodle, & E. Korin (Eds.), *Explorations in critical studies of advertising* (pp. 28–41). Routledge. https://doi.org/10.4324/9781315625768-9

Ong, A. (2021, October 27). This company is making digital humans to serve the Metaverse. *The Verge.* https://www.theverge.com/2021/10/27/22746679/soul-machines-metaverse-digital-humans-labor

Orland, K. (2024, February 27). Meta will start collecting "anonymized" data about Quest headset usage. *Ars Technica.* https://arstechnica.com/gaming/2024/02/meta-will-start-collecting-anonymized-data-about-quest-headset-usage/

Osborne, T., & Jones, P. (2022). Embodied virtual geographies: Linkages between bodies, spaces, and digital environments. *Geography Compass, 16*(6). https://doi.org/10.1111/gec3.12648

Papacharissi, Z. (2011). *A networked self: Identity, community, and culture on social network sites.* Routledge.

Penny, S. (1993). Virtual bodybuilding. *Media Information Australia, 69*(1), 17–22. https://doi.org/10.1177/1329878X9306900105

Pfeuffer, K., Geiger, M. J., Prange, S., Mecke, L., Buschek, D., & Alt, F. (2019). Behavioural biometrics in VR. In *Proceedings of the 2019 CHI conference on human factors in computing systems* (pp. 1–12), paper 110. https://doi.org/10.1145/3290605.3300340

Phillips, A. (2020). *Gamer trouble: Feminist confrontations in digital culture.* New York University Press.

Regalado, A. (2020, August 30). Elon Musk's Neuralink is neuroscience theater. *MIT Technology Review.* https://www.technologyreview.com/2020/08/30/1007786/elon-musks-neuralink-demo-update-neuroscience-theater/

Rinciari, C. (2024). *Fashion brands in the Metaverse: A framework of value creation and engagement through video game collaborations* [Master's Thesis, Politecnico di Milano]. https://hdl.handle.net/10589/219513

Rodriguez, S. (2021, October 30). Facebook's Meta mission was laid out in a 2018 paper declaring "The Metaverse is ours to lose." *CNBC.* https://www.cnbc.com/2021/10/30/facebooks-meta-mission-was-laid-out-in-a-2018-paper-on-the-metaverse.html

Rodriguez, S., & Vanian, J. (2024, October 10). Zuckerberg's metaverse is finally showing signs of life, but it's not from VR. *CNBC.* https://www.cnbc.com/2024/10/10/meta-finally-finds-success-in-ar-vr-three-years-after-changing-name.html

Roy, J. (2024, September 15). Dress to Impress brings new users to Roblox. *The New York Times.* https://www.nytimes.com/2024/09/15/style/dress-to-impress-roblox.html

Saker, M., & Frith, J. (2022). Contiguous identities: The virtual self in the supposed Metaverse. *First Monday, 27*(3). https://doi.org/10.5210/fm.v27i3.12471

Schüll, N. D. (2012). *Addiction by design: Machine gambling in Las Vegas.* Princeton University Press.

Seymour, M., Lovallo, D., Riemer, K., Dennis, A. R., & Yuan, L. (2023). AI with a human face. *Harvard Business Review.* https://hbr.org/2023/03/ai-with-a-human-face

Shakir, U. (2024, October 11). Tesla's Optimus bot makes a scene at the robotaxi event. *The Verge.* https://www.theverge.com/2024/10/10/24267225/tesla-robotaxi-optimus-we-robot

Sidorenko-Bautista, P., Castillo-Abdul, B., Herranz-de-la-Casa, J. M., & Abellán-Hernández, M. (2025). Marketing, advertising, and branding in Fortnite: How do brands and companies connect today to audiences through the metaverse? *Cogent Social Sciences, 11*(1). https://doi.org/10.1080/23311886.2025.2458058

Sito, T. (2013). *Moving innovation: A history of computer moving innovation.* MIT Press.

Soldo, N. (2021, May 31). *The Dubrovnik interviews: Marc Andreessen—Interviewed by a retard.* Fisted by Foucault. Retrieved March 15, 2025, from https://niccolo.substack.com/p/the-dubrov nik-interviews-marc-andreessen

Stempel, J. (2025, April 25). Nike sued over closure of crypto business. *Reuters.* https://www.reuters.com/sustainability/boards-policy-regulation/nike-sued-over-closure-crypto-business-2025-04-25/

Szambolics, J., Malos, S., & Balaban, D. C. (2023). Adolescents' augmented reality filter usage on social media, developmental process, and well-being. *Media and Communication, 11*(4), 129–139. https://doi.org/10.17645/mac.v11i4.7016

Takahashi, D. (2021, June 12). Nvidia CEO Jensen Huang weighs in on the metaverse, blockchain, and chip shortage. *VentureBeat.* https://venturebeat.com/games/nvidia-ceo-jensen-huang-wei ghs-in-on-the-metaverse-blockchain-chip-shortage-arm-deal-and-competition/

Turkle, S. (1995). *Life on the screen: Identity in the age of the internet.* Simon & Schuster.

Webster, A. (2022, May 27). Gucci built a persistent town inside of Roblox. *The Verge.* https://www.theverge.com/2022/5/27/23143404/gucci-town-roblox

Zendle, D., & Cairns, P. (2019). Loot boxes are again linked to problem gambling: Results of a replication study. *PLoS ONE, 14*(3), 1–13. https://doi.org/10.1371/journal.pone.0213194

# Chapter 5
# Whose Metaverse?

**Abstract** In the conclusion to this book, we underscore our reformulation of the term 'metaverse' as a process of metaversification and discuss its implications for spatiality and corporeality. We subsequently reflect on the specific nature of the various companies involved in this process and on the importance of developing alternative narratives that might counter or even subvert hegemonic visions of the metaverse. We also discuss the reality of infrastructures in terms of their material requirements and raise critical questions about the corporate and privatized visions of the metaverse with a specific focus on the ways in which it should be considered a communal and public good.

Rather than focusing on the metaverse as a relatively static and concrete interface or virtual world, this book has sought to critically examine and interrogate the metaverse as a dynamic socio-technical process of worldmaking (Haraway, 2016; Ruberg, 2025). In a nutshell, this process involves the drawing in of increasingly varied types of social activities into virtual worlds while simultaneously pushing the logics of virtualization out into our physical spaces and bodies. On the one hand, multiple forms of labour, ownership, commercial exchange, social interaction, identity, and consumption are being drawn into virtual worlds while opening these up to new forms of commodification, datafication, and exploitation. On the other hand, the companies that create or operate virtual worlds and interfaces are seeking to virtualize and datafy public and private spaces as well as identities and bodies, while simultaneously positioning themselves as tools and infrastructures for the capturing, monitoring, and control of industrial processes and professional labour. In other words, these endeavours can best be understood as a process of metaversification through which various 'tech' companies are seeking to extend their control over spaces and practices both digital and physical.

As both a centrifugal and a centripetal force, metaversification involves three already ongoing and well-established processes. First, virtualization encompasses the creation of virtual versions of physical systems, objects, people, or activities, and the subsequent blurring of the boundaries between physical and virtual worlds. Second, datafication involves the conversion of more and more aspects of human

C. Hesselbein and P. Bory, *Infrastructures of Reality: Metaverse Stories, Spaces, Bodies*,
PoliMI SpringerBriefs, https://doi.org/10.1007/978-3-031-97167-9_5

life and its environment into digital data for quantified analysis or decision making. Third, infrastructuralization consists of making the operation and stability of societies dependent on the functioning of metaversal platforms, technologies, and services. All in all, these processes extend the power and influence of the companies that are engaged in one (or multiple) of these processes over different economic sectors of society as well as various social practices and spheres of life. These three aspects of metaversification are, taken together, likely to profoundly challenge and reshape our understanding of spaces and bodies, and to ultimately reconfigure our reality.

As a means of positioning 'tech' corporations at the centre of society and to justify their ongoing practices and future developments, the metaverse is already here. That is to say, the metaverse is not a mirror or doppelganger world that exists alongside our world. Instead, our world is already being virtualized and only likely to become more so. A great deal of our lives is already spent on the applications, devices, and platforms developed by 'tech' companies regardless of whether these are smartphones or virtual worlds. And a great deal of our lives is already being datafied, sometimes voluntarily, and the future is likely to bring more of the same. These data are being used to build AI models that are subsequently used to create virtual humans and robots, which in turn are supposed to populate and operate in our physical and virtual worlds. The dual processes of virtualization and datafication are already well established and clear even if many of their implications, for better or for worse, still need to be understood. The extent to which the metaverse, or metaverses, will truly become a central infrastructure to our lives and societies remains to be seen, but there is a tremendous amount of resources and power behind the push in this direction. Nevertheless, the contours of this process of infrastructuralization—and its potential links with virtual worlds and devices as well as processes of data collection and automation or roboticization—can already be sketched. This is important to acknowledge in order to anticipate the potential implications of such processes and to provide a starting point for critique and intervention. The end goal of 'tech' companies that are working towards the metaverse is, we believe, to become infrastructures of our reality. These infrastructures facilitate and enable an increasingly wider variety of professional and leisure activities while simultaneously reshaping our understanding of our spaces and our bodies. Needless to say, this raises a number of pressing concerns about whether we, as citizens of this virtualized world, wish to grant responsibility and cede control to such a relatively small and unelected set of corporate actors.

We have chosen to employ the terms 'metaverse' and 'metaversification'—despite significant theoretical and political qualms—because they nonetheless handily remind us of several important aspects. First and foremost, that the metaverse should not necessarily be thought of as a static endpoint but instead as an ongoing process of metaversification that seeks to achieve more goals than those made explicit. It is perhaps precisely for this reason that the metaverse is such a flexible or amorphous and therefore vague term. As an expression of seeming endlessness (Beer, 2024)— in terms of spatiality, creativity, customizability, property, corporeality or identity, and even temporality—the metaverse is employed to justify the creation of a virtualized world while obscuring that this will provide seemingly endless data. Second, the 'metaverse', as implying both transcendence and a new universe, is a term that underscores the grandiosity and hubris of the companies seeking to enact their visions of this new world (somewhat ironically, after Facebook's name change, the metaverse can also be interpreted as a world that is specifically built in the image of Meta). Third, the term 'metaverse' highlights its origins in a range of cultural works that have emerged in the twentieth century, above all sci-fi literature and cinema, that are distinctly utopian as well as dystopian in character. Unfortunately, the narratives put forward in such books and films, as well as the potential implications of the ideas and technologies involved in them, frequently seem to be misinterpreted by CEOs of 'tech' companies who seemingly unironically associate their products with such obvious dystopias. Last, we assert that metaversification is a useful term for highlighting that this process is more than a form of data colonialism and surveillance capitalism or remediation. Although these are all crucial aspects that require acknowledgement, the metaverse is more than media—instead more akin to a virtualized place that is inhabited (but certainly not only a virtual or 'online' world)—and metaverse data signifies an intensification and extensification of data collection on the level of spaces and bodies that promises to be far more all-encompassing and intimate.

Although to some extent all 'Big Tech' companies can be said to be involved in one if not all four of these developments in multiple domains, we have focused on those companies that have affiliated themselves—to greater or lesser extents and in varying degrees—with the development of the metaverse. Not all the companies that we have examined are conventional 'Big Tech' companies. Meta, Apple, and Nvidia are, of course, paradigmatic examples, but companies such as Epic and Roblox are not usually considered as such. Each of these companies has different approaches to the metaverse, sometimes not even employing this term at all, such as in the case of Apple. We have frequently referred to 'Big Tech' companies throughout our chapters, but this is not a term that should be taken as straightforward and stable. Only five years ago, few would have included Nvidia as a member of this group, for example. Perhaps more importantly, we need to be careful—as analysts of communication and media technologies—to not consider the actions of such 'big' companies as well as their smaller siblings as monolithic. In grouping companies together in the 'big' category, the suggestion often seems to be that these companies share similar features, such as being involved in 'tech', as having a high market valuation, and as having a disproportionate influence over technology and society, but the presumption is often also that such companies share a similar ideology, especially if they are based in Silicon Valley. However, there are important differences in the ways that, for example, Apple and Meta address issues of privacy. And although Nvidia and Epic might appear to have similar approaches to virtualization, they are also very different companies with business models and supply chains that are perhaps incomparable. Indeed, the logics of platformization appear not be playing out as smoothly in the game industry as they have in search engines or social media (Chia et al., 2020). The extent to which such differences matter in terms of grouping such companies together as 'Big Tech' or even 'metaverse' companies cannot be discussed here, but it suffices to say that the metaverse is not only a 'Big Tech' endeavour. Indeed, it need not necessarily be an American one, as can be witnessed from the metaverse efforts of companies such as Tencent, Infosys, and Samsung, which operate in completely different socio-cultural and regulatory contexts, and are consequently likely to build virtual worlds that diverge from those created by companies in Europe and the US.

The metaverse—in the sense of virtual worlds where people find distraction or pleasure, engage in creative play or find social connection—need not necessarily be a negative reality. As anyone who has spent any time in virtual worlds can attest, whether they were there for gaming or for socializing, such places reproduce our conventional 'socio-physical' realities along the full gamut, ranging from the most horrible and toxic behaviours to the most profound experiences of beauty and joy. That being said, the crucial question that emerges—when the metaverse is being conceptualized and constructed in the image of a relatively small group of companies—is whether and to what extent the future metaverse can accommodate and reflect the wishes and desires of everybody without their every action and interaction being subjected to extractive processes that first and foremost benefit this relatively small group of 'Big' companies and that are allowing their smaller siblings, such as Epic and Roblox, to grow larger and more powerful. Tellingly, some CEOs are already keenly aware of such concerns, or at least willing to gesture towards allaying such concerns. Mark Zuckerberg, for example, frequently emphasizes that the metaverse will not be built by one single company and that this needs to be done 'together', but never makes clear exactly who this might encompass. Meta, moreover, through its funding of various academic, industrial, and policy partnerships for the 'responsible' development of mixed-reality technologies and worlds, has put itself in a strong position to influence discussions and emergent critiques of the metaverse (Egliston et al., 2024). Furthermore, Tim Sweeney of Epic, has claimed that "if one central company gains control of [the metaverse], they will become more powerful than any government and be a god on Earth" (Ball, 2022, 19), which is a statement reminiscent of similar claims around the development of AI and its supposed omnipotence. And similar to the case of AI, such assertions often appear to shore up and justify the power of the companies that are already in the best position to develop such technologies because new regulations will disproportionly have an impact on smaller and weaker competitors. The tools of regulation and governance initiatives, in other words, are already being challenged and co-opted, illustrating yet another way in which these companies are reshaping reality.

A familiar mantra in Science and Technology Studies is that 'things could be otherwise', which is usually invoked to counter simplistic linear narratives about technological developments or their outcomes, and to underline that such developments involve design choices that can be different. The metaverse, too, can and probably should be otherwise.

At the moment, the most populous, prolific, and productive virtual spaces—both practically and economically speaking—are virtual worlds marked by a strong playful dimension. A game is, by nature, a moment or activity in which we simultaneously discover and construct environments, identities, and relationships (Caillois, 1961; Ortoleva, 2012). However, the sense of playful pleasure or 'playsure' (Vial, 2019) ostensibly enabled by virtual worlds runs the risk of misrepresenting them as neutral conduits for social interaction; gaming and content creation appear as separate activities that do not impact reality. In contrast, the actions of gamers and creators on platforms such as Roblox and Fortnite actively facilitate and reproduce the underlying mechanisms of digital capitalism, thus advancing new forms of commodification and exploitation as well as the datafication of bodies and selves. Virtual worlds, in much of their current state, thus reproduce and exacerbate the broader socioeconomic reality in and through which they have emerged.

The metaverse—or rather the metaverses—are framed in corporate narratives and actions as neutral infrastructures or platforms. Accordingly, metaverse companies neither promote specific ideas of 'being together' nor prescribe ways of relating but only offer specific modalities of interaction and transaction. Yet behind this façade of neutrality, playfulness, and economic opportunity, lies anything but impartiality. In light of this underlying reality, it is legitimate to ask whether activist movements, grassroots communities, or even other companies will be able to appropriate, domesticate, and reshape the metaverse as envisioned by Meta and other dominant players in its current material, socio-cultural, and economic form. More importantly, we must ask whether and how future movements and collectives might acquire the necessary means to challenge the technological, media, and political dominance of these corporate giants. Can different visions of the metaverse truly emerge in a world where the reappropriation of technology—such as happened during the personal computing era of the 1970s and 1980s—now seems virtually impossible? Our current world is one in which even regulators and governments not only frequently support the agendas of companies such as Meta (recent European attempts at regulation notwithstanding) but are also becoming both increasingly incapable of curbing their actions—in part because many 'tech' platforms are supranational—and increasingly dependent on their technologies, infrastructures, data, and services.

It is for this reason that we need new stories before new technologies. If the metaverse was born from the imagination, envisioning a better metaverse will once again require exercising our imagination (Benjamin, 2024). Stories, spaces, and bodies are indivisible elements of what we define as the infrastructures of reality. The stories we carry with us—those we tell and those told to us—shape the conditions for imagining new spaces as much as new forms of embodiment. In short, stories and the imaginations that underpin them carry the seeds for the creation of new realities. At the same time, the remodelling of spaces and bodies enabled by metaversal technologies and worlds can generate and sustain new stories and future-oriented narratives involving individuals, communities, and technological as well as economic actors (McStay, 2023; Mosco, 2023; Turner, 2023). It is precisely for this reason that the construction of virtual worlds should not be left to a small group of dominant companies.

In this book, we have framed the metaverse, in its various technological, spatial, and corporeal dimensions, as an infrastructure of reality. It is equally important, however, to reflect on the reality of infrastructures. As a complex socio-technical ecosystem, the metaverse requires enormous amounts of computational power, energy, and human labour in order to be created, operated, and maintained. This raises the question of whether this cost is sustainable, especially when the metaverse is seen as inseparable from the development of AI. The prevailing scalability model of AI—the belief that increased computational power yields proportionally better performance, particularly for technologies such as Large Language Models—has recently been challenged by the concept of 'diminishing marginal returns' (Maxwell, 2024; Villalobos et al., 2024). According to this view, the incremental benefits of AI systems decrease as more resources are invested. When applied to, or better yet, combined with metaverse technologies, this concept raises a crucial question about the viability and sustainability of such developments and the metaverse in general. Are we truly prepared and can we even afford to invest such vast resources into building this new socio-technical ecosystem? And, more importantly, can we be certain that the ultimate returns, whether marginal or significant, not only augment the quality of immersive experience, but more importantly, also enhance our social, cultural, and political practices and forms of reciprocity in order to improve not just our individual lives, but more critically, our ways of being together? Perhaps above all, how can the metaverse involve and enable commons-based tools, public partic-ipation, and the redistribution of power? As long as we leave the answers to such important questions—issues that have haunted our modern societies for decades if not centuries—in the hands of 'tech' companies, it should not surprise us if their replies reflect their narrow interests rather than our collective wishes. Dreams can swiftly turn into nightmares if we let our reality become their virtuality.

# References

Beer, D. (2024). Extensive culture: Expressions of endlessness in the metaverse and the limits of data accumulation. *Information, Communication & Society, 28*(5), 926–941. https://doi.org/10.1080/1369118X.2024.2413114

Ball, M. (2022). *The Metaverse: And how it will revolutionize everything.* Liveright Publishing Corp.

Benjamin, R. (2024). *Imagination: A manifesto.* W.W. Norton & Company.

Caillois, R. (1961). *Man, play, and games.* University of Illinois Press.

Chia, A., Keogh, B., Leorke, D., & Nicoll, B. (2020). Platformisation in game development. *Internet Policy Review, 9*(4). https://doi.org/10.14763/2020.4.1515

Egliston, B., Carter, M., & Clark, K. E. (2024). Value and virtue in the extended reality (XR) industry. *Information, Communication & Society, 28*(5), 870–889. https://doi.org/10.1080/1369118X.2024.2423339

Haraway, D. (2016). *Staying with the trouble: Making kin in the Chthulucene.* Duke University Press.

Maxwell, F. (2024, November 20). Current AI scaling laws are showing diminishing returns, forcing AI labs to change course. *TechCrunch.* https://techcrunch.com/2024/11/20/ai-scaling-laws-are-showing-diminishing-returns-forcing-ai-labs-to-change-course/

McStay, A. (2023). The Metaverse: Surveillant physics, virtual realist governance, and the missing commons. *Philosophy & Technology, 36*(1). https://doi.org/10.1007/s13347-023-00613-y

Mosco, V. (2023). Into the Metaverse: Technical challenges, social problems, utopian visions, and policy principles. *Javnost—the Public, 30*(2), 161–173. https://doi.org/10.1080/13183222.2023.2200688

Ortoleva, P. (2012). Homo ludicus. The ubiquity of play and its roles in present society. *G|A|M|E Games as Art, Media, Entertainment, 1*(1). https://www.gamejournal.it/homo-ludicus-the-ubiquity-and-roles-of-play-in-present-society/

Ruberg, B. (2025). *How to queer the world: Radical worldbuilding through video games.* NYU Press.

Turner, C. (2023). The Metaverse: Virtual metaphysics, virtual governance, and virtual abundance. *Philosophy & Technology, 36*(4). https://doi.org/10.1007/s13347-023-00666-z

Vial, S. (2019). *Being and the screen: How the digital changes perception.* MIT Press.

Villalobos, P., Ho, A., Sevilla, J., Besiroglu, T., Heim, L., & Hobbhahn, M. (2024). Will we run out of data? Limits of LLM scaling based on human-generated data (Version 2). *arXiv.* https://doi.org/10.48550/ARXIV.2211.04325

The manufacturer's authorised representative in the EU is Springer
Nature Customer Service Centre GmbH, Europaplatz 3, 69115 Heidelberg,
Germany. If you have any concerns regarding our products, please
contact ProductSafety@springernature.com

Printed and bound by CPI Group (UK) Ltd, Croydon, CR0 4YY
28/04/2026
02098543-0001